EXPLORING THE
ISLANDS
of
SCOTLAND

A FRANCES LINCOLN BOOK

First published in the UK in 2008 and the USA in 2009 by
Frances Lincoln Publishers Ltd
4 Torriano Mews
Torriano Avenue
London NW5 2RZ

A catalogue record for this book is available from
the British Library.

ISBN 978-0-7112-2758-3

Printed and bound in Singapore

Produced for Frances Lincoln Publishers Ltd by Julian Holland

Photography and design
Julian Holland

Additional text
Miranda Smith and Denise Stobie

Cartography
Stirling Surveys

Editor
Denise Stobie

Proofreading
Chas Stoddard

Below *On the short crossing across the Sound of Barra, the Caledonian
MacBrayne ferry from north Barra nears the end of its journey to the
new terminal on the island of Eriskay. With the completion of the road
causeway from South Uist to Eriskay and the ferry link to Barra it is
now possible to island hop all the way down the Outer Hebrides.*

EXPLORING THE
ISLANDS
of
SCOTLAND

THE ULTIMATE PRACTICAL GUIDE

JULIAN HOLLAND

F

FRANCES LINCOLN LIMITED
PUBLISHERS
www.franceslincoln.com

CONTENTS

Below *Not a human being in sight! This magnificent stretch of sand at Traigh Losgaintir on the island of Harris is criss-crossed at low tide by several rivers that flow down from the island's mountainous exterior.*

Above *The islands of Scotland are richly endowed with hundreds of important archæological sites. Probably over 3,500 years old, these slabs of old red sandstone are one of many that litter the landscape of Machrie Moor in the southwest of the Isle of Arran. Machrie Moor is littered with hut circles, standing stones and cairns dating back to at least the Bronze Age.*

Above *The islands of Scotland are world renowned for their diversity and abundance of birdlife. Here, on the sea stacks of St Kilda, is the world's largest colony of gannet - during the breeding season over 60,000 pairs nest on the sheer cliff faces of Boreray and the two adjacent stacks of Stac Lee and Stac an Armin.*

INTRODUCTION

Geologically unique, rich in flora and fauna, wild, remote and steeped in history, the many islands around the long and rugged coastline of Scotland are among the most unspoilt and beautiful destinations to be found anywhere on our crowded planet.

Formed during massive and violent upheavals in the Earth's crust nearly three billion years ago and later eroded and shaped by the action of ice, the islands of Scotland also bear the scars of thousands of years of human occupation. From Neolithic settlements, chambered burial tombs, megalithic stone circles and Iron Age brochs, to early Celtic Christian chapels, Viking placenames, clan fortresses, deserted townships of the infamous 'clearances' and more modern relics of both World Wars, the islands of Scotland are a historical treasure trove – second-to-none!

From the sand-blown machair of Tiree and the white shell-sand beaches of Barra to the towering sea cliffs and stacks of remote St Kilda and the dramatic Cuillins of Skye, the Scottish islands are famed worldwide for their beauty. Internationally recognised for their flora and fauna, the islands are also home to many important nature reserves that provide a safe haven for rare and endangered plant and bird species. The surrounding seas, rich in marine life, not only support vast colonies of seabirds but also large numbers of seal, whale, dolphin and porpoise.

Exploring the Islands of Scotland is both beautiful and practical and not only provides the intrepid traveller with a fascinating insight into each island's history and flora and fauna, but also contains useful information on how to get there, tourist information, what to see, where to stay and island walks.

Above *Some of the most beautiful and uncrowded beaches in Europe can be found on the Hebridean islands. Here, in the far southwest of Tiree, its clean white sand washed by Atlantic rollers, Balephuil Bay is one of many fine beaches on that island. Not only is Tiree one of the sunniest places in Britain but it is also one of the windiest.*

FIRTH OF FORTH

BASS ROCK

ISLE OF MAY

Above *After a day's fishing out in the North Sea, these gannets head home across the still waters of the Firth of Forth to their home on Bass Rock. Located just out to sea from North Berwick, Bass Rock is famous for its enormous colony of gannet which, during the breeding season, number around 50,000 pairs.*

BASS ROCK

Dominating the Lothian coastline, the volcanic plug of Bass Rock has had a turbulent past. Owned by the Lauder family for many centuries, its impregnable fortress held out for years against Cromwell's army before being used as a prison for recalcitrant Presbyterian ministers who refused to accept the Government's strict rules on organised religion. Today, it is inhabited by around 50,000 pairs of breeding gannet who leave their mark on the island with vast quantities of white guano.

Firth of Forth

Fidra *Craigleith* *castle* **BASS ROCK** *lighthouse*

North Berwick

A198 *A198*

| 0 | kilometres | 5 |
| 0 | miles | 3 |

HISTORY

First occupied in the 7th century by St Baldred, an early Christian missionary from Ireland, Bass Rock became the property of the Lauder family in the 11th century, remaining in their ownership for many centuries before finally being purchased by Sir Hew Dalrymple in 1706. His successors still own the island today. The ruins of a 15th century chapel dedicated to St Baldred can be seen by visitors to the islasnd today.

The castle, the ruins of which can also be seen today, was built in the early 15th century and during Cromwell's reign of terror, this impregnable fortress held out until 1652, becoming the last royalist stronghold to surrender in Scotland. It was later used as a prison for Presbyterian ministers and their followers who had refused to accept the Government's dictatorial control of the Church. Many of these unfortunate men died in captivity on the island.

Bass Rock became a *cause celebre* for the Jacobites in 1691 when just four prisoners overwhelmed the garrison on the island and held it, with the aid of supporters from the mainland, for three years until they surrendered in 1694.

Following the purchase of Bass Rock by the Dalrymple family in 1706, life on the island returned to some normality. The castle became disused and the island was let to various tenants who made their living from sheep grazing, harvesting gannets and their eggs and fishing. A lighthouse, designed by David Stevenson, was built on the island in 1902 and became automated in 1988.

Today the island is world-renowned for its huge breeding gannet population – numbering nearly 50,000 pairs – who slowly but surely seem to be covering this unique volcanic plug with tons of white guano

NATURAL HISTORY

Apart from its enormous population of breeding gannets, Bass Rock is also home to many other seabirds including guillemot, razorbill, puffin, kittiwake, fulmar and shag. Small numbers of greater black-backed, lesser black-backed and herring gulls have also been known to breed here in recent

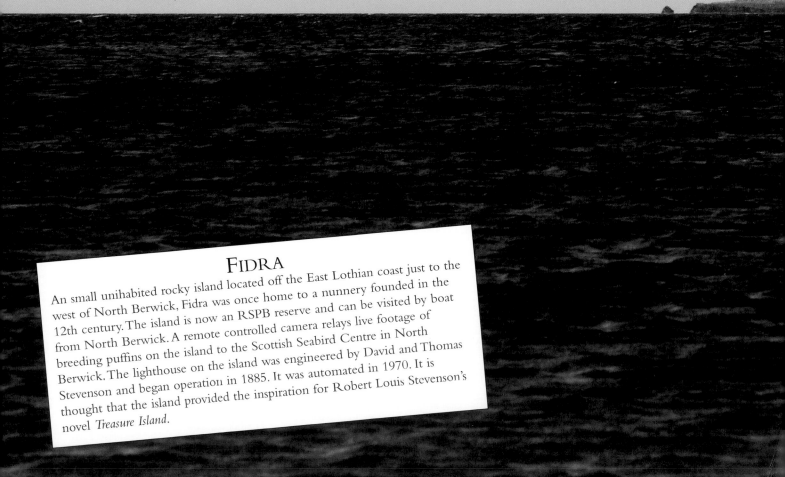

FIDRA

An small unihabited rocky island located off the East Lothian coast just to the west of North Berwick, Fidra was once home to a nunnery founded in the 12th century. The island is now an RSPB reserve and can be visited by boat from North Berwick. A remote controlled camera relays live footage of breeding puffins on the island to the Scottish Seabird Centre in North Berwick. The lighthouse on the island was engineered by David and Thomas Stevenson and began operation in 1885. It was automated in 1970. It is thought that the island provided the inspiration for Robert Louis Stevenson's novel *Treasure Island*.

years. Opened in 2000, the Scottish Seabird Centre in North Berwick offers visitors the chance to view the lives of gannets on Bass Rock and also other bird and sea life on the Isle of May and Fidra through remotely controlled cameras. For more details visit the Centre's website: www.seabird.org

Once used for grazing sheep when the top of the island was covered in grass, little vegetation now remains due to the growing numbers of gannet who have established a noisy and smelly colony during their breeding season. Both the common and Atlantic grey seal are often seen in the waters around the Rock.

HOW TO GET THERE
By sea The *Sula 2* operates boat trips around Bass Rock and Fidra from North Berwick Harbour between April and September. The Scottish Seabird Centre in North Berwick operates photographic boat trips that land on Bass Rock. Places are limited and booking is essential. For more details contact the Centre (tel. 01620 890202) or visit their website: www.seabird.org

ORDNANCE SURVEY MAPS
Landranger 1:50,000 series No. 59

TOURIST INFORMATION
Nearest office: North Berwick Tourist Information Centre, Quality Street, North Berwick, East Lothian EH39 4HJ (tel. 01620 892197) or visit website: www.visitscotland.com

WHERE TO STAY
There is no accommodation on the Bass Rock. However a wide range of accommodation is available in and around North Berwick. For more details contact North Berwick Tourist Information Centre (see above) or visit the town's website: www.northberwick.com

ISLAND WALKS
After landing, visitors to the island can follow a steep concrete path that leads from the old fort past the ruins of St Baldred's Chapel and the lighthouse to the disused foghorn. On a clear day there are panoramic views from the summit across the Firth of Forth, the Isle of May and the Fife and Lothian coastlines.

Left *Visible for miles from the Lothian coastline, the 364ft-high volcanic plug of Bass Rock has an eerie presence with its coating of white guano from the 50,000 pairs of gannet who inhabit the island during the breeding season. Now uninhabited, the island fortress was once a prison for Presbyterian ministers who refused to toe the Government's dictatorial line on religion.*

CRAIGLEITH
Located just off the East Lothian coastline about halfway between Fidra and Bass Rock, Craigleith was heavily fortified for centuries due to its commanding position overlooking the approaches to the Firth of Forth. Until recent years it also had an enormous puffin population during the breeding season but the invasive tree mallow has caused numbers to drop dramatically. Fortunately, thanks to the efforts of SOS Puffin, numbers are now increasing again. New remote controlled cameras operated from the Scottish Seabird Centre in North Berwick give visitors the chance to not only witness live broadcasts from the puffin breeding grounds on the island but of other breeding seabirds such as the cormorant.

ISLE OF MAY

After the martyrdom of St Adrian on the Isle of May by the Vikings in the 9th century, the island became an important place of pilgrimage for hundreds of years. By the 16th century, however, the island had become the haunt of smugglers and wreckers until Scotland's first lighthouse was built here in 1636. Playing an important strategic role during both World Wars, the Isle of May became a National Nature Reserve in 1956 because of its important seabird and grey seal population.

HISTORY

Following major battles between invading Viking armies and the Scots in the 9th century, an order was given by the Danish king, Humber, to slaughter all Christians in Fife. Some, led by the Christian missionary, St Adrian, fled to the Isle of May but were captured and brutally murdered on the island in 875.

Following the martyrdom of St Adrian, the Isle of May became a place of pilgrimage and, in 1145, a priory was built on the island. Over the next two centuries the island changed hands on several occasions, but still remained an important centre for pilgrims, which included members of the Scottish royal family. By the 16th century the island's religious community had waned and was replaced by a small community of fishermen, smugglers and wreckers.

The deliberate wrecking of ships on the Isle of May's rocky coastline was put to an end in 1636 when Scotland's first lighthouse, topped by a burning beacon, was built on the island. The Isle of May was purchased by the Commissioners of Northern Lights in 1815 and a new lighthouse, designed by Robert Stevenson, was commissioned in 1816. The lighthouse became automated in 1989. A further lighthouse was built in 1844 on the northeast coast of the island, but is now out of use.

Due to its strategic position in the Firth of Forth, the Isle of May played an important role during both World Wars. Towards the end of World War I, the waters around the island were scene to one of the most tragic naval blunders of all time, when a large fleet of Royal Navy ships, including battle cruisers and submarines, came into collision with each other and over 100 men were lost. During World War II, the island was an important centre for both anti-submarine detection and radar installations on the vital approaches, via the Firth of Forth, to the Royal Navy dockyard at Rosyth.

NATURAL HISTORY

With its enormous grey seal and breeding seabird population, it is hardly surprising that the Isle of May has been designated as a National Nature Reserve since 1956. From October to January, around 2,000 grey seal pups are born around the island's coastline.

During the spring and summer, the island's precipitous cliff ledges resemble a noisy seabird city with thousands of kittiwake, razorbill and guillemot jostling for space for their nests. Additionally, the cliff edges are riddled with the burrows of the 50,000 pairs of breeding puffin that visit the island each year. Inland, the sky is thick with common, Sandwich and Arctic tern – and beware any human who dares to approach their nesting sites!

HOW TO GET THERE

By sea The *May Princess* operates a daily return passenger ferry service from Anstruther, Fife and the Isle of May between Easter and the end of September. It is recommended to reserve a place on the boat before travelling (tel. 01333 310054). Alternatively, tickets can be purchased from the ticket kiosk at Anstruther harbour. For more details visit Anstruther Pleasure Cruises' website: www.isleofmay.com
Note: Dogs are not allowed on the island.

Above (top) Scotland's first lighthouse was built on the Isle of May in 1636. Due to be demolished after the building of the new lighthouse in 1816 (lower picture) it was saved at the last minute by the intervention of Sir Walter Scott although its height was much reduced to avoid obscuring the new light.

Left *The seas and rocky coastline around the Isle of May support the fourth largest colony of grey seal in northeast Britain. Up to 2,000 pups are born here each winter during which period the island is closed to visitors.*

Left *The puffin is a familiar sight on the Isle of May. During the spring around 50,000 pairs arrive to dig out a short burrow where their single chick is born and raised.*

ORDNANCE SURVEY MAPS
Landranger 1:50,000 series No. 59

TOURIST INFORMATION
Nearest office: Anstruther Tourist Information Centre, Scottish Fisheries Museum, Harbourhead, Anstruther, Fife KY10 3AB. Tel. 01333 311073 (April to October) or visit the website: www.visitscotland.com

WHERE TO STAY
There is no accommodation available on the Isle of May. However, a wide range of accommodation is available in and around Anstruther on the Fife coast. For more details contact Anstruther Tourist Information Centre (see above).

ISLAND WALKS
The only way to get around the Isle of May is on foot, following a series of well marked footpaths that lead to sites of interest and viewing points. All visitors are met on arrival by the resident Warden, who will advise on the best places to visit. Note: Visitors should avoid precipitous cliff edges!

Right *During the breeding season, the precipitous cliff ledges around the coastline of the Isle of May are a raucous home to around 18,000 pairs of guillemot, 2,000 pairs of razorbill and 7,000 pairs of kittiwake. Visitors to the island are warned to be careful on cliff edges. Hard hats are also recommended to protect against dive bombing terns!*

ORKNEY ISLANDS

Mainland/South Ronaldsay

Hoy/Flotta

Shapinsay

Rousay/Wyre/Egilsay

Westray/Eday/Papa Westray

Sanday/Stronsay/North Ronaldsay

ORKNEY MAINLAND

Inhabited for over 6000 years, Orkney Mainland is home to some of the most fascinating archæological and historical sites in these islands. Of particular note, the stone circle known as the Ring of Brodgar and the Neolithic village of Skara Brae should be on any visitor's itinerary. The island is blessed with a diverse range of flora and fauna and is a popular port of call for cruise liners. For the whisky connoisseur, the Highland Distillery should not be overlooked!

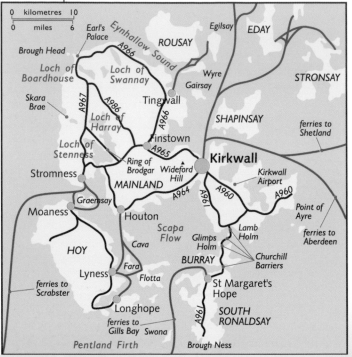

HISTORY

Artefacts in the form of flint tools found in Orkney suggest that the islands have been occupied since at least 6000BC. These early people left no evidence of settlements, but there is an overwhelming number of archæological sites showing that, by 3500BC, the so-called First Orcadians were master builders and stonemasons. Burial cairns such as Maes Howe and Unstan, near Stenness, are considered to be among the finest chambered tombs in western Europe, and were in use for at least three hundred years before being sealed. These and other similar cairns are visible today and can even be entered.

Also in West Mainland, between the lochs of Stenness and Harray, are the breathtaking stone circles of the Ring of Brodgar and the Standing Stones of Stenness, as well as other isolated monoliths. Thought to be astronomical observatories aligned to the Winter Solstice, these outstanding megalithic structures are a fine testament to the skill of the Neolithic Orcadians, and are now part of the 'Heart of Neolithic Orkney' World Heritage Site.

Orkney is also home to the best known Neolithic 'village' in western Europe: Skara Brae. Built entirely of stone, the interconnected houses have stone beds, dressers, seats and, in the middle of each room, even fishponds!

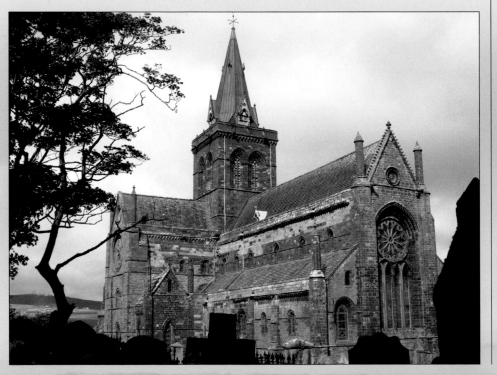

Left *Founded in 1137 by Earl Rognvald-Kali, St Magnus Cathedral dominates the town of Kirkwall. It is dedicated to the martyred Earl Magnus and contains his remains. The building is a mixture of European Romanesque or Norman and early Gothic styles and was considerably extended during the 13th century and late 14th centuries. Following the Scottish Reformation in 1560, the cathedral was used for Protestant worship before becoming a Presbyterian church.*

LORD KITCHENER

Born in Ireland in 1850, Kitchener had an illustrious career in both North and South Africa. In June 1916, as Minister for War in Lloyd George's government, he was to meet secretly with the Russian Tsar. His ship, *HMS Hampshire*, set out in a strong north-easterly gale and tragically struck a German mine - the ship sank within 15 minutes. Of the ship's compliment of 665 men only 12 were saved, and Kitchener's body was never found. The *HMS Hampshire* is now an Official War Grave.

By 700BC, such stone villages were being supplanted by roundhouses and the stonemasons had extended their skills to the working of metal. By the beginning of the 1st century AD, local chieftains were building the huge circular towers known as brochs, of which there are over 100 examples in Orkney. The best of these on Mainland Orkney are in the west, at Gurness and Barwick. Their use began to decline as Orkney fell prey to outside influence from the Romans, Irish and Picts.

Above *The west coast of Mainland Orkney is littered with the remains of wrecked ships, such as this one in Marwick Bay. Perched in a prominent position on the headland in the distance is the memorial to Lord Kitchener, who lost his life nearby on 5 June 1916.*

Right *Overlooking Scapa Flow and founded in 1798, the Highland Park distillery is the most northerly in the British Isles and plays an important part in the Orkney economy. Visitors are most welcome!*

Below *The Grand Princess is one of many cruise liners that call at Kirkwall, boosting the local economy during the summer months.*

Right *Located on a narrow isthmus between the Loch of Stenness and the Loch of Harray, the Ring of Brodgar is one of Orkneys most famous archæological sites. The ring of 27 standing stones (there were originally 60) date from the Neolithic period around 2,500bc and with a diameter of 341ft is the third largest stone circle in the United Kingdom.*

Above *Skaill House is one of the most complete 17th century mansions in Orkney. Located only a short distance from the Neolithic village of Skara Brae, the house is open to the public and also includes two self-catering apartments.*

In the 8th century, a wave of invaders came from the east: the Vikings. At first just isolated raids, probably using the islands as base camps for venturing further afield, by about 800AD the Norsemen were arriving in earnest. It isn't known whether they settled peacefully alongside the native population or conquered it, but for the next 600 years Orkney was an earldom of Norway, an influence still strongly felt today.

Christianity came to Orkney before the Vikings even arrived; there is evidence of Celtic monastic activity at Deerness in East Mainland and at other sites on Mainland, but there is little to see now. Other islands have more concrete evidence of Celtic hermits or papae (fathers). The first Norse Earl to be baptised was Sigurd the Stout in 955. According to the Orkneyinga Saga, a Viking history of Orkney, the capital, Kirkwall, was founded in around 1030 by Earl Rognvald. St Magnus Cathedral was founded in 1137, by which time Kirkwall was a thriving city.

Although most of the other Scottish Islands were ceded to Scotland in 1266, Orkney held out until 1468 because of its Norse heritage – and even then, held on to its own laws. The Udal laws, which gave individual rights to crofters, were ratified by the Scots Parliament in 1567, giving Orkney a unique standing in Scottish history.

In 1486, Kirkwall was created a

Royal Borough and St Magnus Cathedral was given to the corporation of the city, becoming part of the civil community rather than being linked to any specific religious tradition.

In 1581 the illegitimate son of King James V, Robert Stewart, was made Earl of Orkney. Robert and his son, Patrick, were extortionists and tyrants, and are blamed

for the destruction of the Orkney and Shetland lawbooks. Completely dominating the islands in the late 1500s – the family has been accused of 'enslaving' them – these Scots-born lords used forced labour to build their palaces at Kirkwall, Birsay and Scalloway. The Bishop's Palace in Kirkwall, for instance, was acquired by Earl Robert in 1568 and further remodelled by Patrick in 1600. Apparently, Earl Patrick became bored

Above *Now a World Heritage Site and located on the west coast overlooking the Bay of Skaill, Skara Brae is Northern Europe's best preserved Neolithic village. Around 5,000 years old, it was first discovered in 1850 when a major storm uncovered some of its ancient dwellings hidden below the sand dunes. The impressive remains of this village, complete with original furniture, were fully excavated in the late 1920s by Professor Gordon Childe. Adjoining the site is a visitor centre.*

with the project and acquired land through having the previous owner tried and executed on false theft charges, and there, in 1607, built the Earl's Palace. This is now held to be the finest Renaissance building in Scotland. The Stewarts held sway for 40 years but were the last true Earls, as their excesses led to greater scrutiny and more control from the Scottish government. Patrick and his son, another Robert who rose in revolt,

Right *At the far northwest tip of Orkney Mainland are the impressive ruins of Earl Robert's Palace. It was built on the site of an older Bishop's Palace in 1569 by Earl Robert Stewart, the illegitimate son of King James V of Scotland. By the early 18th century, however, the palace had fallen into disrepair.*

were finally executed in 1615.

A possibly ancient custom survives in the capital Kirkwall, on Christmas Day and New Year's Day: the Kirkwall Ba' Game. This wild 'football' match is played in the streets of the city between the Uppies and Doonies, and which team is played for is often determined by whether the player was born 'Up the gates' or 'Doon the gates'. Although the present form of the game seems to have been determined in about 1800, it is thought to be much older. It used to be played on Old Christmas Day (January 6) and Old New Year's Day. The Gergorian calendar was only adopted in Orkney in the mid 1800s.

No trip to Orkney would be complete without a visit to the Highland Park Distillery,

the most northerly in Scotland. Founded in 1798, it was originally the site of an illegal distillery run by one Mansie Euson - a church officer who kept some of his stock in the pulpit!

NATURAL HISTORY

Orkney is home to a large number of wildlife reserves, with six RSPB Reserves and one Scottish Wildlife Trust Reserve on the mainland

Left *Much of the northwestern part of Mainland Orkney is fertile farmland, with cattle raising forming a major part of the island's economy.*

alone. Hobbister RSPB Reserve, for instance – also designated as a National Scenic Area – has fine moorland scenery where the keen birdwatcher can see hen harrier, merlin, peregrine and short-eared owl, as well as the elusive red-throated diver. Puffin breed between May and August at the Brough of Birsay near Marwick, and several pairs of rare pintail can also be seen.

Mute swan are particularly prominent on the lochs of Harray and Stenness, and can be observed from the Bridge of Brodgar. Greylag geese can also be observed here; they first bred on the Mainland in 1985, prompting the theory that they were Icelandic pairs who liked the scenery so much that they decided to stay!

Orkney is renowned for its overwintering migrant birds. For instance, in December 1990 17,000 migrant waterfowl were counted on the loch of Harray. In autumn, the arrival of pochard, tufted duck, goldeneye and whooper swan can be observed. The swans used to stay for just a few days, but some pairs now appear to stay all winter.

Orkney's sheltered areas of shallow sea are excellent habitats for many species of waterfowl, such as shag, cormorant, the resident eider, great northern diver, velvet scoter and, in winter, Slavonian grebe. At Kirkwall and Stromness, Iceland and glaucous gull can regularly be seen, particularly at sewage outfalls.

Surveys have shown that Orkney is a real birdwatcher's paradise, with 28% of the UK population of purple sandpiper and 25% of the population of curlew to be found here. The sharp-eyed birdwatcher can also see the occasional osprey, long-eared owl, gyrfalcon and white-tailed sea eagle.

The birdlife on Orkney is both diverse and seasonal, so for information on what is currently resident on the islands contact the local RSPB office (tel. 01956 850176). Please note: many species have their own unique, local Orkney names, so you may have to ask for a translation!

Mammals to be seen on (and off) Orkney include dolphin, porpoise, several species of whale and the very shy otter, best seen at dusk and dawn if one is patient. Common and grey seal are also easily spotted, often basking on the rocks around the coastline. Inland, the Orkney vole is a fascinating creature. Resembling the field vole, but larger and with shorter and paler fur, this species originated in the Balkans but was introduced to Orkney at least 4,000 years ago - quite how or why is a mystery. They are becoming much more elusive due to modern farming practices and the hunting by local birds of prey.

Above *Cut off at high tide from the Mainland and with its curiously slanting slabs of sandstone exposed at low tide, the Brough of Birsay contains many important archæological sites, including the ruins of a Pictish village and the remains of Earl Thorfinn's Palace and his Great Christ Church, monastic buildings and Norse longhouses. The Brough can be reached on foot at low tide from the Point of Buckquoy on the northwest tip of the Mainland.*

HOW TO GET THERE
By air Loganair operate flghts from the major British airports to Kirkwall Airport. For further details contact Loganair (tel. 01856 872494/873457} or visit their website: www.loganair.co.uk
By sea Northlink Ferries operate a vehicle and passenger ferry service between Aberdeen and Kirkwall. For further details contact Northlink Ferries (tel. 0845 6000 449) or visit their website:

www.northlinkferries.co.uk

In the summer months, John O'Groats Ferries operate a passenger ferry between John O'Groats and Burwick on South Ronaldsay. For further details tel. 01955 611353 or visit their website: www.jogferry.co.uk

Pentland Ferries operate a vehicle and passenger ferry between Gills Bay on the Scottish mainland and St Margaret's Hope on South Ronaldsay. For further details tel. 01856 831226 or visit their website: www.pentlandferries.co.uk

ORDNANCE SURVEY MAPS
Landranger 1:50,000 series No. 6 & 7

TOURIST INFORMATION
Nearest office: Kirkwall Tourist Information Centre, 6 Broad Street, Kirkwall, Orkney, KW15 1NX (tel. 01856 872856) or visit www.visitorkney.com

WHERE TO STAY
There is a wide range of accommodation on the Orkney Mainland. For more details contact the Kirkwall Tourist Information Centre (see above).

ISLAND WALKS
With an area of 202 square miles, and the sixth largest Scottish island, the Orkney Mainland is by far the largest of the Orkney group of islands and, as such, offers many opportunities for walkers. There is far too little room in this book to describe all of the ideal walks but some of these include the walks around the Point of Ness, Binscarth Woodland, the Marwick Circular and the Swartland Drovers Road. For more details of these walks visit the website: www.walkorkney.co.uk

Details of many other walks on Mainland Orkney can be obtained from the Kirkwall Tourist Information Centre,

Above and right *Scapa Flow had been the main base for the Royal Navy's Grand Fleet since the beginning of World War I. However, during 1939, at the beginning of World War II, a German U-boat penetrated the defences and sank HMS Royal Oak with the loss of 833 lives. This disaster spurred Winston Churchill to order the building of what became known as the Churchill Barriers, forming an impregnable defence network around Scapa Flow. At the peak of their construction thousands of workers, including 1,300 Italian prisoners of war, were employed. Using pre-cast concrete blocks, the barriers and associated block ships (above) linked the Orkney Mainland with the islands of Lamb Holm, Glimps Holm, Burray and South Ronaldsay.*

6 Broad Street, Kirkwall, Orkney, KW15 1NX (tel. 01856 872856) or visit www.visitorkney.com

SOUTH RONALDSAY

This island covers more than 5000 years of history in a relatively small acreage. The Tomb of the Eagles, a recently excavated chambered cairn, dates from about 3150BC and was in use for about 800 years. Its name comes from the fact that, as well as human bones, remains of the talons of at least 10 sea-eagles were also found. The remains of the Liberian ship *Irene* can also be seen here, whose distress calls were responsible for the launch and subsequent loss of the lifeboat, *TGB,* in 1969. South Ronaldsay is joined to Burray by the Ayre of Cara, the only man-made ayre in Orkney: an 'ayre' being a spit of land enclosing a lagoon which is open to the sea.

Above *St Margaret's Hope is the largest settlement on the island of Southy Ronaldsay. A regular vehicle and passenger ferry operates between here and Gills Bay on the Scottish mainland.*

THE ITALIAN CHAPEL

This beautiful creation, born of faith and ingenuity, was constructed by the Italian POWs of Camp 60 on Lamb Holm between 1942 and 1945, using the few materials available to them – mainly concrete and plasterboard. Two Nissen huts had been built together to create one large building, with the intention that one should be a school and the other a chapel. Eventually, the school was abandoned and the raw materials were used to build an altar, rood screen and statues of incredible delicacy, as well as the impressive frontage. Frescoes were mainly painted by Domenico Chiocchetti, based on a card his mother gave him when he left home for the war. The chapel remains one of the most poignant legacies of the Second World War.

HMS ROYAL OAK

On 14 October 1939, following orders for an immediate assault on the British fleet, a German U-boat penetrated the defences of Scapa Flow. Commanded by Lieutenant Gunther Dean, *U-47* found the anchorages almost empty and gave the order to sink the battleship *HMS Royal Oak*, one of only two ships there. The ship went down in 13 minutes. Of 1400 men on board, 833 died, either going down with the ship or killed by spilled fuel oil.

Memorial and wreath-laying ceremonies are held each year on the tragedy's anniversary, with RN divers raising a White Ensign on the propellor shaft of the sunken vessel.

HOY

ferry to Scrabster
Stromness
MAINLAND
Hoy Sound
Moaness GRAEMSAY
St John's Head
Ward Hill
Houton
Scapa Flow
B9047
Old Man of Hoy
Rackwick
CAVA
Oil Terminal
HOY
FARA
Lyness
FLOTTA
Longhope
South Walls
SWONA
Pentland Firth
0 kilometres 10
0 miles 6

The second largest island in the Orkneys, Hoy is steeped in history and affords the visitor some of the most spectacular scenery in these islands. The cliffs alone are known as some of the best in Britain. From the magnificent Old Man of Hoy to the naval shipwrecks at Scapa Flow, Hoy is a veritable treasure trove for the discerning traveller.

Below *The island of Hoy is so named from the Old Norse word 'Haey' which translates rightly so as 'High Island'. The second largest island in Orkney, it also has the highest peak. Ward Hill, towering some 1,571ft above sea level and here seen shrouded in mist. The Royal Society for the Protection of Birds owns nearly 10,000 acres of moorland and sea cliffs in the north of the island.*

HISTORY

The history of Hoy and South Walls – two islands treated as one for census purposes – stretches back more than 5,000 years. The Neolithic 'Dwarfie Stane', arguably the only rock-cut chamber tomb in Britain, and the Iron Age Skeo Broch are testimony to early habitation.

Norse occupation is attested by the tale that the 10th century King of Norway, Olaf Tryggvason, forced Sigurd 'the Stout', Earl of Orkney, to become Christian – on pain of death. Sigurd submitted to baptism on Hoy, but apparently did so in name only.

Hoy is also home to the most northerly Martello Towers in Britain, built as defences against US privateers during the Napoleonic Wars.

The island's major historical legacy, however, lies in the 20th century. The two World Wars saw this island housing the main naval base for northern Britain at Lyness, overlooking Scapa Flow. In mid-November 1918, the majority of the German fleet, some 74 ships, was ordered here as part of the Armistice agreement. In the six months that followed, conditions aboard the vessels broke down and plans were laid to scuttle the fleet. Due to a breakdown in communications as to the state of the Armistice talks, the German admiral in command, Ludwig von Reuter, believing that a state of war still existed, gave the order *Paragraph eleven*, which signalled the final demise of the fleet. Remarkably, only nine lives were lost in this action, the last of WWI

The military history of this bay is so broad that one needs to do one's own research: visit the website www.scapaflow.co.uk for further information. At one point, 30,000 men were stationed at Lyness – whereas today, there are only 200 inhabitants on the whole island. Two new piers were built here in WWII: one is now home to a colony of tern and the other to tysties, otherwise known as black guillemot!

NATURAL HISTORY

Hoy's natural history is so diverse that there are too many national

BETTY CORRIGALL

One of the more poignant stories in Hoy's history relates to poor Betty Corrigal. Although dates differ (18th or 19th century, depending on the source), the story seems to be the same. The young girl became pregnant by a sailor, the shame of which drove her to attempt suicide by drowning: fortunately (or unfortunately, as it turned out), she was saved from this. Soon afterwards, having been thwarted in her first attempt, she hanged herself instead. She was buried on the lonely, windswept parish boundary. The grave was rediscovered in the 1930s and, during WWII, soldiers tidied it up and built a fence around it. The present tombstone was designed by a local artist.

classifications to list definitively! There is an RSPB Reserve in the north of the island, a Scottish Wildlife Trust (SWT) Reserve in South Walls, a National Scenic Area and finally, Hoy is designated a Special Area of Conservation (SAC).

The RSPB Reserve includes the famous 'Old Man of Hoy', a rock stack eroded by the sea and possibly destined to disappear altogether, if comparisons are made between the stack today and drawings from last century. A mixture of cliff and moorland, the reserve offers visitors views of great skua (the second largest breeding colony in the UK), red grouse, golden plover and cunlin on the moors and guillemot, razorbill and kittiwake on the cliffs. It is also the site of the most northerly natural woodland in Britain.

The SWT reserve was first established to protect a plantation of *primula scotica*, the local species which was the original emblem of the Trust.

For more information on the wildlife of Hoy, visit www.rspb.org.uk/scotland or, for the SAC, tinyurl.com/2hs5x9

HOW TO GET THERE
By sea A passenger and cycle ferry operates from Stromness on the Orkney Mainland to the island of Graemsay and Moaness Pier on North Hoy. For more details tel. 01856 850624.

A vehicle and passenger ferry operates from Flotta on the Mainland to Lyness on Hoy. For more details tel. 01856 811397.

ORDNANCE SURVEY MAP
Landranger 1:50,000 series No. 7

Above *The world-famous Old Man of Hoy is a 450ft-high red sandstone sea stack on the dramatic northwest coast of Hoy. It was first climbed as recently as 1966 during a live televised ascent by a team which included Chris Bonington. Access to the cliffs overlooking the Old Man is via a sometimes very muddy cliff and moorland footpath that starts at the car park at Rackwick.*

TOURIST INFORMATION

Nearest office: Kirkwall Tourist Information Centre, 6 Broad Street, Kirkwall, Orkney, KW15 1NX (tel. 01856 872856) or visit www.visitorkney.com

WHERE TO STAY

There are several hotels, bed and breakfast and self-catering establishments on Hoy. For more details contact the Kirkwall Tourist Information Centre (see above).

ISLAND WALKS

Hoy is certainly the wildest and most rugged island in Orkney. Its remote interior offers many walking opportunities for hillwalkers, but sensible clothing and footwear should always be worn. Hoy's west coast offers some of most stunning cliff walking in the UK but, again, care should be taken and a watchful eye kept on the weather.

One of the most exciting walks is from the Moaness ferry pier following the narrow road up the dramatic Rackwick Valley. From Rackwick, there is an often muddy footpath that follows the cliffs and then across moorland to the famous Old Man of Hoy. Another footpath runs from Rackwick up through the Glens of Kinnaird, skirting Hoy's highest point, Ward Hill (1,571ft), before finally arriving back at Moaness. Berriedale Wood, close to the Glens of Kinnaird, are one of the few locations in Orkney that still contain an area of the original woodland that once covered the islands.

In the southern part of the island, a track leads from Heldale to the summit of Binga Fea (505ft) and on to the island's largest loch, Heldale Water. The outfall from the loch flows down the Burn of Greenheads and ends in a dramatic waterfall on the west coast.

Below *The beautifully cared-for naval cemetery at Lyness not only contains the graves of many British and Commonwealth sailors who perished during the two world wars, but also those of German sailors whose bodies were washed up around the shores of Hoy including this one of Johannes Thill, who perished on 7 December 1918. Many famous incidents in the history of the Royal Navy are recorded on the gravestones, including the Battle of Jutland, the sinking of* HMS Hampshire *and the explosion of* HMS Vanguard, *all in World War I, and the torpedoing by a German U-boat of* HMS Royal Oak *in Scapa Flow in 1939.*

HMS VANGUARD

In July 1917, the St Vincent class battleship, *HMS Vanguard*, exploded while at anchor inside Scapa Flow. 843 men were lost – all but 2 of the crew. The disaster, which led to the redesign of British warships, was thought to have been caused by an undetected fire in a coal bunker, which ignited the cordite in one of the two munition magazines that served the turrets amidships. On board of the *Vanguard* at the time was a liaison officer from the Imperial Japanese Navy, Commander Kyosuke Eto.

The wreck of the ship is now an Official War Grave.

FLOTTA

Lying between Hoy and South Ronaldsay in the mouth of Scapa Flow, the previously quiet farming community of Flotta found itself, in the early years of the 20th century, strategically vital to the Royal Navy. Everything changed for the inhabitants at the start of WWI, when the original population of only 431 in 1910 was boosted by the influx of military personnel to the extent that a WWI photograph shows a boxing match with an audience of 10,000!

It took until 1970 for fresh water to be piped to Flotta – today, 10% of Britain's oil exports are piped through the oil terminal there. Built in the mid 1970s to process oil from the North Sea oilfields, the terminal has positively affected the economy of the whole of the Orkneys without massive negative effects on the environment or community of any other island, as the technological development has been deliberately confined to Flotta.

THE LONGHOPE LIFEBOAT TRAGEDY

On 17 March 1969, the Liberian freighter *Irene* got into difficulties off Grimness, on South Ronaldsay. The boat's distress calls alerted the lifeboats of Kirkwall on the Mainland and Longhope on Hoy, both of which launched immediately into the teeth of a very strong south-easterly gale.

Longhope's lifeboat, the *TGB* (her name thought to have been derived from the initials of her anonymous sponsor), was tragically capsized off South Ronaldsay by a wave estimated to have been around 100ft high. The entire crew of eight were lost, which included both first and second coxswains and their respective sons.

A bronze statue was erected in Osmondwall Cemetery to honour both the disaster and the courageous crew.

Above *Located at Brims on South Walls, Longhope Lifeboat Station opened in 1834 and has saved over 500 lives. The station is now a museum. For more information tel. 01856 701332.*

SHAPINSAY

O ne of the most fertile Orkney islands, Shapinsay has two nature reserves, but only one village! The island's agricultural heritage, stretching back thousands of years, was transformed by the Balfour family in the 19th century. Shapinsay is supposedly the island from which the Orkneys were 'conquered' by the Romans. There is no historical evidence for this suggestion, but it remains an interesting legend.

HISTORY

Archæologically, Shapinsay remains one of the least excavated islands in the Orkneys. Many tantalising mounds probably hide rich finds, especially given the history of other islands. Mor Stein, a standing stone which looks like a 10ft hand, is said to have been thrown by a giant after his fleeing wife - fortunately for her, he apparently missed.

Shapinsay has several Iron Age brochs, Burroughness being one of the best examples in the whole of the Orkneys. At the same time as the brochs were being built, Roman history states that General Agricola and his fleet 'conquered' Orkney in 43AD. The fleet apparently landed on Shapinsay, and Roman artefacts have been found at broch sites – probably acquired through trade. Roman historians have never been noted for their accuracy, although they were superb spin doctors!

During later Norse occupancy of the islands, Haakon of Norway's fleet is said to have mustered at Elwick Bay before the battle of Largs in 1263. Although the outcome of the battle was inconclusive, Scotland eventually won the war with the signing of the Treaty of Perth in 1266. The Orkneys, however - always independent - continued to be Norwegian for another 200 years.

Shapinsay's history owes much to the Balfour family. In the 1840s, David Balfour is known for transforming the island's farming techniques, but before this the family was renowned for its staunch Jacobite sympathies and, as a result, their house was burned down by government soldiers in 1746. Enough money trickled into the family through later marriages to allow them to build Cliffdale, later extended to become Balfour Castle.

Traditional farming practices, first introduced on Shapinsay in the Neolithic period, were completely reformed after 1846 when David Balfour instituted massive agricultural change. Former small farms were divided into 10-acre fields across the entire island and, by 1874, arable farming had increased from 748 acres to over 6,000.

The last Balfour in line died in 1960 – having had 4 wives and no heirs. Captain Tadeusz Zawadski, a Polish cavalry officer, with his Scottish wife and her children, then took over the castle. The captain had walked across Europe in order to reach Britain in 1941, was posted to Orkney and fell in love with the people…and the free fishing. The couple now run the main hotel on the island.

Shapinsay is also noted for being home to the last person ever to be executed in

Above *As part of the planned development of the Balfour Estate on Shapinsay, Balfour village was built in the late 18th century to provide homes for the craftsmen who were employed on the estate. This ornate round building is all that remains of a gas works that was built to supply the estate in the mid-19th century.*

BALFOUR CASTLE

Balfour Castle was designed around an existing late 18th century house for David Balfour by Edinburgh architect David Bryce as the Orkney summer home of the Balfour family. It was completed in 1848 and incorporated seven turrets, 12 exterior rooms, 52 rooms and 365 panes of glass. The last of the Balfours, David Hubert Ligonier Balfour, died in 1961 leaving no heirs and the castle was sold to former Polish cavalry officer Captain Tadeusz Zawadski and his family. The castle is now run as a luxury hotel and, along with its gardens, is also open to the public between May and September on Sunday afternoons. Booking for these guided tours is essential. For more details contact Balfour Castle (tel. 01856 711282) or visit their website: www.balfourcastle.com.

Above *The gatehouse of Balfour Castle is the main entrance to the most northerly castle hotel in the world!*

Orkney, an island woman accused of 'child murder'.

There are WWII coastal battery emplacements at Salt Ness, which can still be seen today.

NATURAL HISTORY

Shapinsay is rich in wildlife, especially for the dedicated birdwatcher. There is an artificial loch at the RSPB Mill Dean Reserve, where one can see pintail, wigeon, shoveller, pochard and water rail during the breeding season, whooper swan in winter, and greylag goose all year round. Hen harrier and merlin are also regularly sighted.

The south east corner of Shapinsay is a 46-acre Scottish Wildlife Trust Reserve. Mainly maritime heath, it supports breeding gull, tern, skua and wader, and the abundant wildflowers are spectacular throughout the year.

HOW TO GET THERE

By sea Orkney Ferries operate a regular vehicle and passenger ferry between Kirkwall and Shapinsay. For more details contact Orkney Ferries (tel. 01856 872044) or visit their website: www.orkneyferries.co.uk

ORDNANCE SURVEY MAPS

Landranger 1:50,000 series No. 6

TOURIST INFORMATION

Nearest office: Kirkwall Tourist Information Centre, 6 Broad Street, Kirkwall, Orkney, KW15 1NX (tel. 01856 872856), or visit the website www.visitorkney.com

WHERE TO STAY

In addition to luxury accommodation at Balfour Castle (see box on left),

Right *One of the more unusual additions to the Balfour village was this sea-flushing toilet, which is located on the harbour's edge. Across the harbour is another eccentric 19th century building - The Douche - which was used as a salt-water shower and has a roof that was a dovecot!*

Shapinsay offers a small choice of quality bed and breakfast and self-catering establishments. For more details contact Kirkwall Tourist Information Centre (see above).

ISLAND WALKS

With little traffic and a good network of lanes and tracks, little Shapinsay is ideal for the walker. The coastline is certainly worth exploring, especially along the fine sandy Bay of Sangarth, the rock formations at the Foot of Shapinsay and try and discover Odin's Stone - apparently associated with the Norse god 'Odin' - which is a large block found at the western end of the beach around Veantrow Bay. The northwestern tip of Shapinsay, known as The Galt, is also good for walking and spotting birds and seals.

Left *Shapinsay contains many archæological sites including this solitary megalithic standing stone known as Mor Stein, located at the eastern end of the island.*

ROUSAY

With at least 166 identified archæological sites on these three islands, Rousay and neighbouring Egilsay and Wyre have been quite rightly dubbed 'Egypt of the North'. Every footstep takes the visitor back into the rich history of Orkney – and into the sheer variety of Orkney wildlife. Rousay and its satellite islands have a plethora of RSPB reserves, so birdwatchers and casual visitors alike can enjoy themselves – and take the wonderful sunsets as a bonus!

Above *Taversoe Tuick is one of the many chambered cairns that can be visited on Rousay. Apart from a similar circular cairn on nearby Eday, it is unique in having two storeys that had separate entrances and were not connected to one another. Today, visitors can descend to the lower chamber by way of a ladder.*

WYRE

This tiny, spear-shaped island is home to Cubbie Roo's Castle, built around 1150 and one of the first in Scotland to have a square central keep. 'Cubbie Roo' is a derivation of the Viking name Kolbein Hruga, a formidable man by all accounts. The castle was defensive; the family dwelling was probably at nearby Bu Farm. St Mary's Chapel, next to the castle, is of the same date and was probably the family church.

Wyre is excellent for birdwatching, with a diverse range of habitats. Seal can regularly be seen on the shore.

HISTORY

Known as the 'Egypt of the North' – over 100 archæological sites have been identified – and a Site of Special Scientific Interest, Rousay has been inhabited for over 5,000 years. From neolithic chambered tombs to a 19th century manor house, visitors to this peaceful island can see remains from every era of Orkney's history.

The most impressive site is the Midhowe chambered tomb on the west coast, now contained within a large protective building, and the nearby Iron Age broch with walls still standing over 4 metres high.

At Moaness, a Viking graveyard revealed the grave of a woman and infant, with the famous Westness brooch, and two boat burials. The boats' occupants were buried with tools and weapons, and four arrowheads were found in the skeleton of one.

Probably the oldest 2-storey house in Orkney is the Tofts, dating from at least 1600. Now in disrepair, it is the result of 'clearances', Rousay being the only island in the Orkneys to suffer from this policy. In the mid-19th Century, George William Traill evicted 210 people as he thought sheep would be more profitable than growing corn. Having been bequeathed the Westness estate along with its considerable debts, Traill's nephew, General Burroughs, raised rents on the island and, in 1873, commissioned the building of Trumland House. The rent rise, and subsequent evictions after a Royal Commission enquiry,

earned Burroughs a reputation as one of the worst landlords in Orkney. Today the gardens of Trumland House are open to the public, and the house is being renovated after a 1985 fire.

NATURAL HISTORY

Rousay is a birdwatchers' paradise. The west coast's maritime heath supports large breeding colonies of birds, including arctic tern, great and arctic skua, oyster-catcher, golden and ringed plover and snipe. Rousay's plant life in this area includes *primula scotica*, the Scottish primrose, unique to the north coast and islands of Scotland, grass of parmassus, and spring squill.

Hellia Spur has colonies of guillemot, razorbill, fulmar and kittiwake, while puffin nest in burrows at the top of the cliffs. Waterfowl of many kinds can be seen at Wasbister Loch and, inland, hen harrier, merlin, peregrine and red-throated diver can be spotted.

One of the few wooded areas in Orkney, Trumland Wood, supports woodland bird species such as willow warbler, song-thrush and dunnock. Winter and late spring are the best time for keen birdwatchers to see migrant species such as long-tailed duck and great northern diver.

The Trumland RSPB Reserve is well-signposted, with a circular walk in the centre of the island.

HOW TO GET THERE

By sea: Orkney Ferries run a regular service using the *MV Eynhallow* between Tingwall on Orkney Mainland and

Rousay, Egilsay and Wyre, between May and September. Please note that advanced booking for cars is required before 17.00 on the previous day, and Egilsay and Wyre are request stops! To book, and for further details, tel. 01856 751360 or visit the website www.visitrousay.co.uk

ORDNANCE SURVEY MAPS
Landranger 1:50,000 series No 2

TOURIST INFORMATION
Nearest office: Kirkwall Tourist Information Centre, 6 Broad Street, Kirkwall, Orkney, KW15 1NX (tel. 01856 872856) or visit www.visitorkney.com or contact Mark and Dianne Hull (Rousay Tourist Association) (tel. 01856 821225).

WHERE TO STAY
There is a hotel and several self-catering and bed and breakfast establishments on Rousay. To book, contact Rousay Tourist Association (see above). There is also a hostel and campsite at Trumland Farm (tel. 01856 821314).

ISLAND WALKS
Westness Walk – 'the most important archæological mile in Scotland' – covers Rousay's history from 3500BC to the 19th century in the time it takes to walk from Westness Farm, about 4 miles from the ferry pier, to Midhowe Cairn.

Faraclett Walk includes Bigland round cairn and the remains of Rinyo neolithic settlement, as well as spectacular views over Westray Firth.

There are several guided walks on the many RSPB reserves on the islands (tel. 01856 821395). The warden is permanently based on Egilsay, but is prepared to travel!

Below *The island of Rousay as seen from the nearby Orkney Mainland. Separating them is the treacherous Eynhallow Sound with its swirling currents and rip tides.*

Above *Dramatically located on Rousay's southwest coast, Midhowe Broch dates back over 2,000 years to the Iron Age. Protected on one side by the sea and by ditches and a rampart on the landward side, it certainly was an impregnable building. This massive structure was built using enormous flag stones and came fitted with its own internal fresh-water spring.*

Below *Adjacent to Midhowe Broch on the island of Rousay is the famous Midhowe Chambered Cairn. Over 5,000 years old, it was originally excavated in the 1930s by the local landowner and found to contain 12 compartments with the bones of 25 people. Over 75ft long, it is now roofed over to protect its fragile state.*

EGILSAY
Egilsay is the permanent home of the RSPB warden in this group of three islands. This reserve was set up specifically to create and preserve the best possible environment for the rare and elusive corncrake, 8 pairs of which were recorded on the island in 1997.

Egilsay is also famous as the place where St Magnus was murdered. St Magnus' church, with its round tower, can still be seen today. Dating from the second quarter of the 12th century, it was probably built on the site of an earlier church.

WESTRAY

As imaginatively named by the Vikings as Sanday – Westray means Western Island in Old Norse – it is much nicer to use the modern epithet of 'Queen o' the Isles'. The second largest of the North Isles and comparable to St Kilda as a haven for birds, Westray is rich in archæological remains and a paradise for birdwatchers. It is also home to much of Orkney's fishing fleet and a huge exporter of cattle and sheep.

HISTORY

Archæological remains on Westray show evidence of continuous occupation since around 3000BC. Chambered tombs and the remains of a Skara Brae-type settlement go back to the Neolithic Age, while Bronze Age burnt mounds and Iron Age brochs are a testament to later population of the island.

In the sand dunes north and west of Pierowall, 9th century Viking graves have been found: weapons, jewellery and combs all helped to date the excavation. These show that Westray was one of the earliest Norse settlements in the Orkneys, probably because Pierowall is one of the safest harbours in the entire archipelago. At Quoygrew, on the north side of Rackwick, a series of structures stretching inland from the shore for about 55 yards represents a continuous occupation of over 1000 years! Finally abandoned in 1937, these buildings are essential to the understanding of the historical change in Orkney.

Noltland Castle, the most impressive building on the island, seems to have been built in around 1560. Its founder, Gilbert Balfour, was a notorious conspirator in the plots of Mary, Queen of Scots and, having fled Scotland, continued his nefarious habits in Scandinavia. He was executed in Sweden in 1576.

Today, Westray is known for its fine cattle and sheep exports and for being the only Orkney island involved in cod, rather than herring, fishing. Westraymen are still held to be among the best fishermen in Scotland and the Isles.

NATURAL HISTORY

The RSPB reserve at Noup Head, the most northerly 2 miles of cliff on the west side of Westray, is second only to St Kilda as 'seabird city'! In May and June, keen ornithologists can see auk, kittiwake, guillemot, black guillemot, razorbill, fulmar and puffin, amongst others. A recent colony of gannet have also begun nesting on the island. Visitors are warned not to disturb the arctic tern,

Above *Located on the northwest tip of Westray, Noup Head lighthouse was built in 1898 by David Stevenson. The surrounding cliffs are an important RSPB reserve with the largest population of breeding seabirds in the Orkneys.*

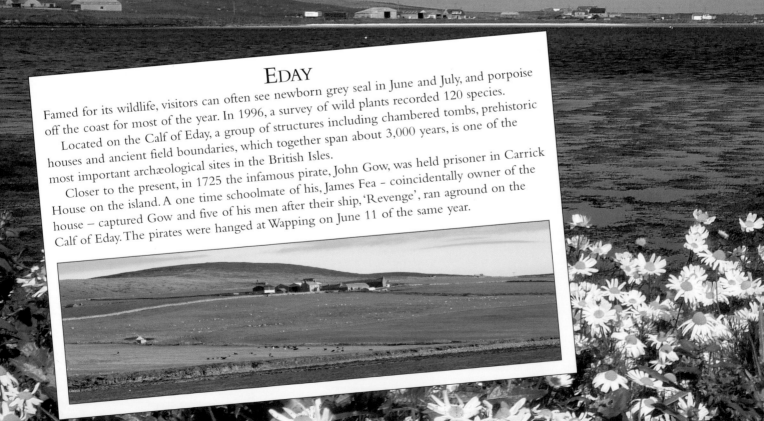

EDAY

Famed for its wildlife, visitors can often see newborn grey seal in June and July, and porpoise off the coast for most of the year. In 1996, a survey of wild plants recorded 120 species.

Located on the Calf of Eday, a group of structures including chambered tombs, prehistoric houses and ancient field boundaries, which together span about 3,000 years, is one of the most important archæological sites in the British Isles.

Closer to the present, in 1725 the infamous pirate, John Gow, was held prisoner in Carrick House on the island. A one time schoolmate of his, James Fea – coincidentally owner of the house – captured Gow and five of his men after their ship, 'Revenge', ran aground on the Calf of Eday. The pirates were hanged at Wapping on June 11 of the same year.

as they can be nasty!

Elsewhere, maritime heath habitat on Westray can be spectacular in summer, with thrift, parnassus, bird's foot trefoil, sea campion and spring squill blooming in profusion from April to late August.

Seal, whale and otter can be seen by the patient observer, but be warned – they are elusive.

HOW TO GET THERE
By sea Orkney ferries run a regular service between Kirkwall and Westray from May to September. There is also one afternoon ferry a day from Kirkwall to Papa Westray. Advanced booking is required for cars, preferably before 1600 on the day before travel. For further details tel. 01856 872044 or visit the website www.orkneyferries.co.uk
By air The world's shortest scheduled flight runs daily between Westray and Papa Westray. It takes longer to board and disembark than it does to fly between islands! Otherwise, Loganair run daily flights between Kirkwall and Westray. For further details tel. 01856 872494/873457, or visit the website www.loganair.co.uk

ORDNANCE SURVEY MAPS
Landranger 1:50,000 series No. 5

Below Pierowall is the main community on Westray and is also the departure point for the short ferry crossing to Papa Westray. Its natural harbour was favoured by the Vikings and, during July of each year, is the scene of the Westray Regatta.

TOURIST INFORMATION
Nearest office: Kirkwall Tourist Information Centre, 6 Broad Street, Kirkwall, Orkney, KW15 1NX (tel. 01856 872856), or visit the website www.visitorkney.com

WHERE TO STAY
There is a good choice of accommodation on Westray. For more details contact Kirkwall Tourist Information Centre (see above).

ISLAND WALKS
Westray offers some of the best cliff walks in the Orkneys. The West Westray Coast Trail starts at the car park at East Kirbest and ends just over 5 miles further on at Noup Head. Other walks include the Noup Head Loop, Tuquoy and Mae Sand and Castle O'Burrian and the Bay of Tafts. A leaflet describing these beautiful walks is available from the Kirkwall Tourist Information Centre.

Above Noltland Castle was built by the ruthless and disliked Gilbert Balfour in the mid 16th century. Balfour was appointed Sherriff of Orkney by Mary Queen of Scots but fled to Sweden in 1572, where he was executed for conspiracy four years later.

Above Built in the 12th century by a Norseman, Cross Kirk stands in a stunning position on the shore of the Bay of Tuquoy. The church's graveyard remains in use and contains many fine gravestones. The remains of a Norse longhouse have also been discovered nearby.

PAPA WESTRAY
Visitors to the Orkneys can get confused here, as Papa Westray and Papay are the same island – so be careful!

People have been farming on Papa Westray since before the Egyptians built the pyramids, and the island is home to the oldest existing house in northern Europe, at the Knap of Howar. It is also known for being the one of the oldest Christian sites in Scotland: there are chapels dedicated to St Tredwell and St Boniface on the island, both of whom lived in the 8th century. St Tredwell's (actually Triduana's) Chapel is one of the most visited pilgrimage sites in Orkney. The word 'pap', when found in Norse placenames, refers to the Celtic Christian priests or 'fathers' who often lived as hermits on remote islands. There is a nature reserve at North Hill on Papay, managed by the people of the island in conjunction with the RSPB and Scottish National Heritage.

SANDAY

Sanday, an unimaginative Old Norse name meaning 'Sandy Island', has been inhabited for at least 6,000 years. Every era of its history has its own visible remains, from the Stone Age to the present. Sanday's wide, sandy beaches – for which it is named – and widely diverse wildlife make it a perfect destination for both naturalists and walkers. Today, Sanday is noted for its beef cattle and fishing industry.

HISTORY

Archæological remains on Sanday show that it has been continuously inhabited since 4000BC and that at some periods has been mostly underwater. There are several neolithic tombs and Iron Age brochs on the island. Artefacts show that the Vikings arrived in the 8th century; a boat burial with remains of a man in his 30s, a woman of about 70 and a 10-year-old child were found in a recent excavation at Scar, along with several unique finds including a superb carved whalebone plaque. There is also a submerged forest at Otterswick.

Many ancient chapel sites exist, often with newer buildings on top. The abandoned Cross Kirk is the best example.

Sanday is notorious for shipwrecks, and it is said that ministers used to pray for a ship to run aground – so the island could have a supply of firewood! The first lighthouse was finished in 1806 at Start Point, housing the first revolving apparatus in Scotland. The present lighthouse was built in 1870 by Thomas Smith and Robert Stevenson, and became automatic in 1962.

At Lopness Bay, the remains of the WWI German destroyer *B98* can be seen; while at Letton there are concrete buildings belonging to a WWII Chain Home Low radar station.

Today, Sanday is noted for its beef cattle and local fishing industry, and tourism remains a major factor in the local economy.

NATURAL HISTORY

With its wide, sweeping sandy beaches, extensive sand dunes and machair, lochs and marches, Sanday is a walker's and birdwatcher's paradise. It is particularly attractive to wading birds, both breeding and migrant. In spring redshank, turnstone, dunlin, ringed and grey plover and many other species pass through on their way to their northern breeding grounds. Bar-tailed godwit and knot are particularly worth looking for, as they

Above *Sanday was a hive of military activity during WWII. In addition to a reserve radar station, a dummy airfield was also built to distract German bombers from attacking Scapa Flow. The operations centre, known as the 'Brickie Hut' still stands on the edge of Cata Sands.*

Below *The sweeping crescent of Scuthvie Bay is located at the far eastern tip of Sanday. A short distance to the east is Start Point lighthouse which was engineered for the Northern Lighthouse Board by Robert Stevenson. When it became operational in 1806 it was the first Scottish lighthouse to have a revolving light. Start Point was automated in 1962 and is now powered by a bank of 36 solar panels.*

STRONSAY

Low-lying and with wonderful sandy beaches, Stronsay prides itself on being one of the friendliest Orkney Isles.

Visitors should take care, however – Stronsay was home to the famous 'Stronsay Beast', washed ashore in 1808. Possibly a sea serpent, but later dismissed as a basking shark (though at least 10 ft longer than any known basking shark ever recorded), 19th century zoologists argued for years over its origin. Even today, the verdict as to its species is 'not proven'.

If you should be unfortunate enough to come across the beast, simply visit the Well of Kildinguie. It has a reputation for being able to cure all known ailments except the Black Death. Presumably residents tried…and failed.

will be changing into their breeding plumage.

In late summer, one can also see little stint and curlew sandpiper. Start Point, in the east, is a particularly good spot to watch for migrants. After storms, rarities can often be seen after they have been blown off-course.

Without intensive farming, many habitats have been sustained for Sanday's flora. Lady's bedstraw, grass of parmassus and several species of buttercup, eyebright and orchid are just some of the species which make Sanday so colourful in spring and summer.

Sanday has the biggest common seal population in Scotland: over 4% of the total UK population! Grey seal also breed here, and both otter and whale can be seen offshore – if you watch carefully.

HOW TO GET THERE
By sea Orkney Ferries run a regular service between Kirkwall, Eday and Sanday from May to September. Vehicles should be booked in by 1600 the day before for some sailings, and bookings are made at the Kirkwall office (tel. 01856 872044), or visit the website www.orkneyferries.co.uk
By air Loganair run a regular service to Sanday from Kirkwall. For details and reservations tel. 01856 872494/873457 or visit the website www.loganair.co.uk

ORDNANCE SURVEY MAPS
Landranger 1:50,000 series No. 5

TOURIST INFORMATION
Kirkwall Visitor Information Centre, 6 Broad Street, Kirkwall, Orkney, KW15

Above *During the scuttling of the German High Seas Fleet in Scapa Flow on 21 June 1919, not all of the ships were successfully sunk. One of these, the destroyer B98, was subsequently towed away but as she was passing Sanday, the tow line broke and she ran aground in the Bay of Lopness. Today, her turbines and boilers are still clearly visible from the beach at low tide.*

1NX (tel. 01856 872856) or visit the website www.visitorkney.com

WHERE TO STAY
There is an extensive range of accommodation on Sanday, from hotels and bed and breakfast establishments to self-catering and caravans. There is also a hostel and campsite.

ISLAND WALKS

Whether you visit Sanday just to get away from it all, or as a botanist, ornithologist or historian, the best way to explore the island is on foot. For walkers, the best way to see the island is to divide it into three separate sections.

The first is the eastern peninsula that starts at the Bay of Newark and continues along the shore of the Bay of Lopness to Start Point and its lighthouse. Heading north, follow the shoreline of the Bay of Scuthvie to Tofts Ness and then along the shoreline of Otters Wick.

The second section is to follow the northerly peninsula by skirting around the sands at the outlet of Lamaness Firth up to Whitemill Bay, then follow the coastline in a westerly direction before ending up at the Bay of Brough.

The final section is to explore the long southwesterly peninsula, following the rugged west coast from the Bay of Brough all the way down to Spur Ness, before returning through the dunes to Backaskaill Bay and the harbour village of Kettletoft. Fortunately for the thirsty walker, there are two public houses in the village!

Right *One of three enormous wind turbines that were erected near the ferry terminal at Spur Ness came into operation in 2004. The wind farm produces a total of 8.25MW and will provide electricity for the islanders for the next 20 years.*

NORTH RONALDSAY

The most northerly of the Orkney Isles, this island retains many of the older customs of the archipelago, and is one of the last places to follow the practice of communal farming. The unique Ronaldsay sheep spend most of the year grazing on seaweed washed up on the beaches, kept there by a wall which surrounds the whole island (about 13 miles). At lambing and shearing times, every farmer on the island herds the sheep into 'punds': stone-built collecting areas.

Worth visiting is the Stan Stane, a lone standing stone unusual in having a hole drilled through the top. It may have been used as a 'sighting stone' for an astronomically-based stone circle, but local legend has it that a giant woman stuck her finger through a stone on the beach and carried it to its present location.

Above and left *Located on the Els Ness peninsula the Quoyness chambered cairn is one of the prime archæological sites on Sanday. It dates back to the Neolithic period, around 5,000 years ago. It is open to the public and entry is only possible by crawling through a 20ft long tunnel into a high central chamber. It is thought that when the cairn was built Els Ness was then an island.*

Below *Kettletoft is the main harbour on Sanday and, until the new ferry terminal was built at Spur Ness in the 1990s, was the main entry point for visitors to Sanday. The harbour grew up around the fishing industry which reached its peak during the herring boom in the late 1880s and early 1900s.*

Above *Sanday is an ornithologists paradise. Along with large numbers of greylag geese, seabirds and waders that visit the island, Sanday is also home to the short-eared owl, here seen above perched on a fence post. Short-eared owls usually mate for life and the males exhibit a vigorous courtship display during the breeding season. Although they generally hunt at night they can often be seen looking for prey during daylight hours by swooping low over fields and grassland.*

Below *Waiting for the fish to bite as the sun goes down over Petta Water in the remote heart of Shetland Mainland. The main road that runs alongside this stretch of water passes through a glaciated valley overlooked by the Mid Kame and East Kame range of hills. In the distance, beyond Mid Kame, is the summit known as Gruti Field (902ft) with its solitary Hag Mark Stone.*

SHETLAND ISLANDS

MAINLAND

YELL

UNST

FETLAR

WHALSAY

FAIR ISLE

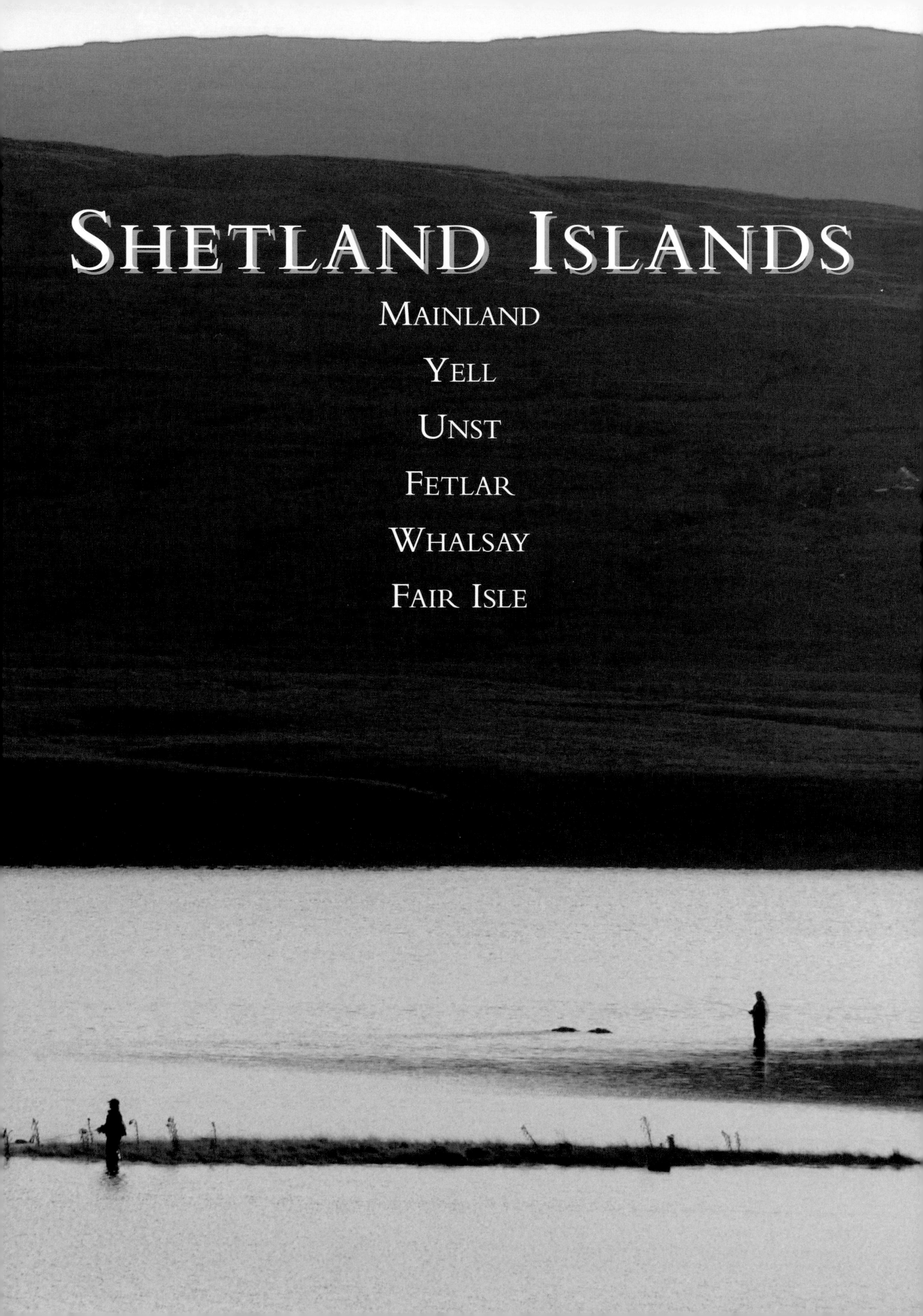

SHETLAND MAINLAND

Shetland Mainland has something for everyone. Walking, fishing, birdwatching, sailing, whale watching – the list is endless for the visitor. Shetland's capital, Lerwick – closer to the Arctic Circle than to London, and 93 miles from the Scottish mainland to the south – is here. More than 17,000 people, 80% of the islands' population, live on Shetland Mainland.

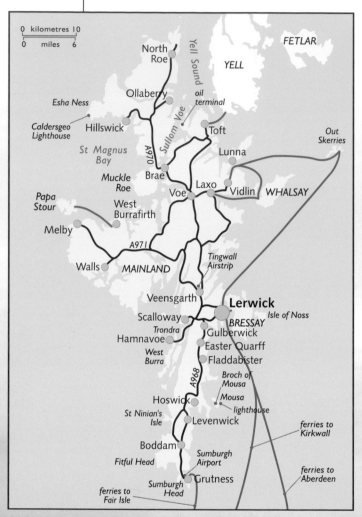

HISTORY

This area is rich in history and remains from Neolithic times to the modern day, and many fantastic discoveries have been made. Stanydale Temple in the west mainland is a Neolithic settlement with remains of houses and field boundaries, probably dating from 2500–2000BC, when the first Neolithic farmers settled in Shetland. In 1905, on the southern end of Dunrossness peninsula, a storm revealed part of what turned out to a prehistoric village. The site had been occupied in Neolithic times, and the Bronze Age houses are still intact – in fact a whole village is there, including primitive streets, stone walls and cooking hearths. Later structures have also been unveiled, including an Iron age broch, a wheelhouse, Pictish houses and a medieval Norse village. A longhouse was built there by the first Norse settlers, and the stone foundation has survived. Jarlshof also has the 16th century Old House of Sumburgh, built for Robert Stewart who became Lord of the Northern Isles in 1564. The name Jarlshof was invented by Sir Walter Scott for the medieval farmhouse in his book The Pirate, but is now used for the whole site.

Below *Once the medieval capital of Shetland, Scalloway and its ruined castle lie only a few miles west of the present capital of Lerwick. Recent improvements to the harbour and the construction of a marina have guaranteed the future prosperity of this important centre of the fishing industry.*

PAPA STOUR

This is a low-lying island off the west coast of Shetland Mainland. Only around 30 people live here, although people have lived here for hundreds of years – the remains of a 12th century Norse wooden farmhouse found at Biggings are unique in the Shetlands. It was also the site of an early Christian monastery – Papa from the Norse papae meaning 'priests', and Papa Stour meaning 'big island of the priests'. In the 18th century, the island was used to isolate lepers, although they were not actually lepers, only suffering from a hereditary and disfiguring skin disease. The island is a Special Area of Marine Conservation.

Above *Located in a stategic position on a peninsula overlooking Scalloway harbour, Scalloway Castle was built in 1600 by the ruthless Patrick Stewart, Earl of Orkney and Shetland. Inhabited for only a short period, the castle soon fell into disrepair and is now open to the public.*

On a small island on the Loch of Clickimin just north of Lerwick, there is an oval structure called Clickimin Broch, which was excavated in the 1950s. This very well-preserved Iron Age stone tower dates from 700BC. Later, a large stone dyke was built right around Clickimin, followed by a blockhouse and in the late Iron Age, a wheelhouse.

There are many evidences of Pictish habitation as well. In 1993, the Mail Stone was found in Mail. This has a Pictish warrior with battleaxe and club carved on it. In 1958, a hoard of Pictish silver was found under a slab in the remains of a monastery on St Ninian's Isle. St Ninian was the first missionary to visit northern Scotland and died in 422AD after establishing the monastery. It is thought that the silver may have been buried to hide it from the Vikings when they began their raids on the islands.

The medieval capital of Shetland was Scalloway, the name meaning the 'bay of the booths' In 1600, Earl Patrick Stewart built Scalloway Castle, at the expense of the local people, after succeeding his father as Earl of Orkney and Lord of Shetland in 1593. The Earl moved the law assembly, the Lawting, from Tingwall to Scalloway, to Law Ting Holm at the north end of the loch. Patrick Stewart became immensely wealthy because of the taxes and confiscations of property that he imposed but in 1615, he and his son were executed because of the excesses

Above *Dore Holm is a magnificent natural arch of red sandstone located just off the spectacular coastline of the Esha Ness peninsula in the northwest of the Mainland.*

he had carried out in Shetland.

Bressay Sound is one of the best natural harbours in Scotland, so over the centuries it has been the focus of many different foreign invaders and traders. During the same period, Lerwick has grown from a small fishing village to become the centre for a healthy fishing industry. – the name Lerwick is Norse for 'muddy bay'. The Vikings and Dutch traders have been followed by Scots, English, Russians and many others. Hillswick has Shetland's oldest public house, The Booth, dating from 1698, named because a Hamburg merchant,

TRONDRA, EAST AND WEST BURRA

One of the Scalloway Islands, Trondra is home to the Shetland Croft Trail, where one can see various breeds of Shetland livestock, discover the art of crofting and learn about traditional crafts, such as boatbuilding. From the fishing village of Hamnavoe on West Burra, a short walk leads to one of the best beaches (and picnic spots) in Shetland, at the Sands of Meal. Further along the coast, the lighthouse on Fugla Ness affords stunning views out to sea.

At Papil is the kirk of St Laurence, which is famous for the discovery of three early Christian carved stones – two of these can be see at the Shetland Museum in Lerwick.

In 1970, the three islands were connected to the mainland by a series of bridges, which has helped reverse the decline in population and led to an increase in tourism for these islands.

Above *Get away from it all! An idyllic scene looking across West Voe from Duncansclate on West Burra towards East Burra.*

SHETLAND PONIES

Shetland ponies have wandered across the islands since at least the time of the Vikings. Known as shelties, they are very hardy, well adapted to the harsh climate. Because they are so tough they have been used for work and transport since the earliest times. The locals even used their long tails, manes and shaggy winter coats for fishing lines and nets. In the 19th century, when the Mines Act of 1847 stopped the use of children in the mines, the Marquis of Londonderry leased the island of Noss to rear Shetland ponies for use in his coal mines. The Shetland Pony Stud Book Society was founded in 1890 to 'promote the breeding of these ponies' and maintain their height at less than 10.2 hands.

Above *Traditional crofting methods are still widely practised in the Shetlands. Here a small field of hay ricks near Laxo would have been harvested in the traditional way by hand using a scythe and rake.*

Adolf Westerman, had a trading booth there in 1684. To service the Dutch traders and the herring fleets, the shoreline at Lerwick was made into a waterfront of warehouses in the Middle Ages and some of these survive to this day. They are called 'lodberries', which were piers built by the merchants to load and unload goods from the warehouses.

The waters round the Shetland Mainland have seen many confrontations. In 1640, Spanish ships attacked the four Dutch frigates which were there to defend the fishing fleet, sinking two of them. In 1653, Admiral Monk was sent with a fleet by Cromwell to attack the Dutch fleet commanded by Van Tromp. The skirmish was unresolved, but the English landed at Lerwick and remained there.

In 1665, a fort was built in Lerwick by Charles II's master mason, John Mylne, to defend the island against Holland during the Second Dutch War. It was attacked and burned by the Dutch only eight years later in 1673. Renovated in 1782, it was named Fort Charlotte after the wife of George III. In 1674, a battle between a Dutch frigate and the Royal Naval frigate Newcastle took place at at Ronas Voe. The English triumphed and many of the Dutch are buried at Hollanders' Knowe

Lerwick was also the place where one of Shetland's most famous people was born in 1792. Arthur Anderson joined a partnership that became the Peninsular and Oriental Steam Navigation Co (P&O). This made him very rich, and until very recently, P&O ran ferry services to Orkney and Shetland.

By the mid-19th century, Lerwick's population had grown to 3,000 and it had become the main administrative centre for the mainland. The fishing industry still provided for the community, with the village of Voe becoming an important centre for cod fishing, while on the north side of Olna

Firth, a whaling station operated from 1904 to 1928.

Although Scalloway was now not the isalnd's capital, from 1942 to 1945, it was the base for what has become known as the 'Shetland Bus'. A vital link of small fishing boats was operated out of Scalloway, making clandestine missions across the North Sea to Norway, under the noses of the Germans, carrying saboteurs, munitions and equipment to help the Norwegian resistance.

Today, at Sullum Voe, there is the largest oil and liquefied gas terminal in Europe, which processes about one million barrels a day, and is one of the main industries for the Shetlands. Construction of the terminal was begun in 1973 and completed in 1982, providing employment for more than 6,000 local people in construction alone. It also has its own airstrip.

The fishing and oil industries hold sway on the mainland. Sailing skiffs called Shetland Maids gracefully sail in and out of Lerwick harbour and the town is thriving. One of the traditional customs maintained there is worth noting – Up Helly Aa (Old Yule), is held on the last Tuesday of January. It is a dramatic fire festival that developed from Shetland guizing or 'dressing up'. The locals build and burn a galley, in true Viking fashion, after a magnificent procession through the streets.

NATURAL HISTORY
Shetland Mainland has lochs and sea, moorlands and sand. It is a paradise for wildlife, with breeding birds and otters,

Below *The evening ferry from Laxo heads up Dury Voe to the island of Whalsay. All of the inhabited outer islands of Shetland are well served by regular ferry services with excellent fare concessions not only to locals but also visitors - especially senior citizens!*

while whales swim past offshore. The landscape is very varied, with old red sandstone, ancient gneiss and schist and fertile volcanic soils. There is seal, puffin, black guillemot and raven, and waders inland and on its shores, as well as many wild flowers beside the many small lochs and streams. The north of the island has the most amazing carved cliffs, stacks, natural arches and caves, including the uninhabited island of Uyea and the jagged Ramna Stacks.

The south end of the island is a long peninsula that runs 25 miles south from Lerwick and is a mixture of moorland and farmland. The rock is old red sandstone, which is about 370 milllion years old, and the area also has many sand dunes. The RSPB reserve at Sumburgh Head is a wonderful place to see puffin (in Shetland dialect, 'tammie nories') from May until mid-August most years. It is also one of the best places to spot killer and other whales. Shetland's first lighthouse was built there by Robert Stevenson in 1814. Nearby, at the Loch of Spiggie, is an RSPB wildfowl reserve, where birdwatchers can see migrating

Above *The remote Lunna Ness peninsula on the Mainland east coast is the location of the tiny 18th century Lunna Kirk and of Lunna House. Built in the 17th century, the latter was used as a headquarters by Norwegian resistance fighters during World War II.*

whooper swan and greylag geese.

In the centre of the island, there is Tingwall, a fertile valley north of Scalloway. This is the best place to see many of Shetland's native wildflowers, including species of orchid. White Ness takes its name from the limestone outcrops in the area. In summer, on the Hill of Sound, there are many moorland species of breeding birds, including curlew and whimbrel. The head of Weisdale Voe is home to many wading birds, while the Loch of Girlsta in the east, is, one of the biggest and deepest lochs in the Shetlands, famous for Arctic char and brown trout. It is said that the Loch was named after Geirhilda, daughter of Hrafna Floki, who drowned there in about 870 while on they were on their way to Iceland.

In West Mainland, Da Wastside is a

BRESSAY AND NOSS
Lying just off the east coast of Mainland, near Lerwick. this small old red sandstone island is 7 miles long and 3 miles across. It has been very important to the fishing industry since the days of the Dutch traders. It is also the source of Bressay stone, which was quarried from Aith Ness in the north and used for building. From Ward Hill (742 ft high) there are fine panoramic views, and the high cliffs at the Ord have natural arches. The little island of Noss off its eastern shores was once joined to Bressay by a thin strip of sand, but now is separate. Noss is a National Nature Reserve. The eastern cliffs are great for birdwatchers – they can see breeding seabirds, including great skua and Arctic skua.

MOUSA BROCH

On Mousa, off the southeast side of Shetland Mainland, there is one of the finest surviving examples of an Iron Age fortress unique to Scotland. Built from the flagstone found on the shore nearby, the broch, or tower, is a drystone construction built without mortar. It has a hollow walls, now home to thousands of stormy petrel, with mural galleries that allow access to the space between the two walls. It is one of more than 500 brochs, but is considered to be one of the best, and the only one that is complete. It is 50ft in diameter and has a height of 43½ft.

peninsula that reaches out into the Atlantic. It is littered with lochs where otter and migrating and wintering wildfowl and waders can be seen. In the Skeld and Deepdale areas, gneiss on the north coast and granite on the southeast have resulted in dramatic cliff scenery. At Bixter, the geology changes as Walls Boundary Fault leads into a landscape of old red sandstone with granite intrusions. Bixter Firth is a sheltered sea loch and very popular with birdwatchers and trout fishermen.

On the North Mainland, the Northmavine cliffs at Eshaness have spectacular views, and this part of Shetland Mainland provides some of the best hill and coastal walking. Most of the east side of this area is schist and gneiss. Ronas Hill and the west side is pink granite. Eshaness is black basalt volcanic rocks and old red sandstone, and at Urafirth, there is salt marsh where you can find oyster plants.

At the westernmost tip of North Mainland. there is a narrow seam of dark material – in a stream at the corner of the seam, there are tiny pieces of wood, peat particles and clay that are thought to be at least 40,000 years old – the remains of coniferous woodland that would have been there at that time. In these fossils is evidence of oak and a Mediterranean species of heather that indicate the climate was much milder at the time.

Eshaness is a good place during the bird migration season, and one of best places to see whale and dolphin in Yell Sound. There is a collapsed cave, the Devil's Cave, near the lighthouse there.

Ronas (Red Ness) is the most remote and only mountainous part of Shetland Mainland. Red granite has formed cliffs at Ronas Voe, and the nearby Lang Ayre is one of the loveliest, if remote, beaches. Ronas Hill is the highest point, and has alpine plants, although it is often shrouded in mist. The island of Uyea is formed of rocks of Lewissian gneiss, some of the most ancient in the world, and is joined to the Mainland by a sandy causeway.

HOW TO GET THERE

By sea Northlink Ferries operate a daily vehicle and passenger ferry service between Aberdeen and Lerwick. On alternate days this service also calls at Kirkwall in Orkney. For more details contact Northlink Ferries (tel. 01856 885500) or visit their website: www.northlinkferries.co.uk

By air Loganair operate direct flights to Sumburgh Airport from Heathrow,

Above *Located on the southwest outskirts of Lerwick, Clickimin Broch is a fine example of an Iron Age stone tower now much reduced in height. Originally built on an island in the surrounding loch, it was connected to the shore by a causeway.*

Birmingham, Manchester, Glasgow, Edinburgh, Aberdeen, Inverness and Orkney. For more details contact Loganair (tel. 0345 222111) or visit their website: www.loganair.co.uk

ORDNANCE SURVEY MAPS
Landranger 1:50,000 series Nos. 1, 2, 3 & 4

TOURIST INFORMATION
Nearest offices: Lerwick Tourist Information Centre, Market Cross, Lerwick, Shetland, ZE1 0LU (tel. 08701 999440) or visit www.visitshetland.com or Sumburgh Tourist Information Centre, Wilsness Terminal, Sumburgh, Shetland, ZE3 9JP (tel. 08701 999440).

WHERE TO STAY
There is a wide range of accommodation on Shetland Mainland. For more details

Below *During the summer months, Lerwick Harbour is a regular port of call for foreign cruise liners. With the island of Bressay in the background, an Italian liner with a smile on its bow leaves Lerwick for its next port of call.*

contact Lerwick Tourist Information Centre (see above)

ISLAND WALKS
The Shetland Amenity Trust organise guided walks at many stunning locations around the Mainland, including Keen of Hamar National Nature Reserve, Hermaness National Nature Reserve, Isle of Noss, Huxter, Sandness – Dale of Walls, Hoswick – Mosquito Memorial, Sumburgh Head and Mousa. For more details contact either the North Shetland Ranger (tel. 01957 711528) or South Shetland Ranger (tel. 01595 694688).

Above *Founded in the 17th century as a port for the herring fishing industry, Lerwick is now the administrative centre of Shetland with a population of just under 7,000. Its busy harbour is alive with the comings and goings of fishing boats, ferries and support vessels for the offshore oil industry.*

Below *In the far south of the Mainland Sumburgh Airport is the main entry point for visitors to the Shetlands. Not only does it handle passenger flights to and from the UK but is also a vital base for helicopters flying to North Sea oil platforms.*

YELL AND UNST

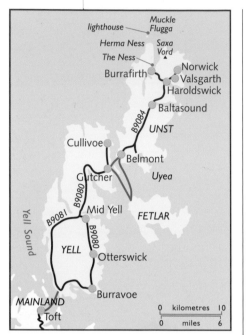

Yell and Unst, with Fetlar, are the North Isles of Shetland. Least visited by tourists, they actually have much to offer the birdwatcher and naturalist. Yell has around 1,000 inhabitants: many live a crofter's existence, and the nearby oil industry also provides jobs. Unst is the most northerly inhabited island of Shetland.

HISTORY

The Vikings made their mark on these islands. On the shore of Kirk Loch on Yell is the ruined Kirk of Ness, dedicated to St Olaf. Local legend says this marks the spot from which Leif Eriksson set out on his famous expedition in 1001AD.

On the southeast tip of Unst is the ruin of the 16th-century Muness Castle, built in 1598 by Lawrence Bruce. He was a rogue who cheated the locals by altering the weights and measures system to his advantage. The castle burned down in 1627.

During the following years, the islanders prospered. In the 19th century, however, the 'clearances' began, and many of the crofters were forced off the land.

The sea around Yell regularly claimed victims. At Gloup, a stone woman and child stare out to sea as a memorial to 58 men who drowned in 1881, when a storm wrecked six fishing-boats. However, the island also provided safety – during the First World War, German U-boats often sheltered in Whale Firth.

Today, salmon farming is taking over as an industry on both islands, while Baltasound on Unst - once the centre of the herring industry - has an airstrip used by the oil industry, and is the port for the export of talc.

NATURAL HISTORY

Yell is famous for its otter population, which are often seen on the northeast coast and near the breakwaters at Ulsta. Common and grey seal, as well as porpoise, also swim there.

Most of Shetland's eight species of orchid are found on the moorland, while rare plants such as butterwort and sundew like the deep peat. The moors are also home to curlew, plover, red-throated diver and merlin. At the Lumbister RSPB Reserve in the northwest, there is some of the most varied coastal scenery, where many different seabirds nest and breed.

The interior of Unst has a wide variety of habitats and there are two National Nature Reserves: Hermaness and the Keen of Hamar. Hermaness combines coastal scenery and wild moorland, with more than 100,000 nesting seabirds, including gannet, puffin and black guillemot. The moorland is home to the largest colony of great skua found anywhere in the world. There are at least 800 pairs, who dive-bomb anyone who goes near their nesting grounds!

The Keen of Hamar is 74 acres of serpentine rock, which breaks into tiny fragments that give the landscape a strange lunar appearance. It is famous for the plants that grow there, including Edmondston's Chickweed, discovered in 1837 by 12-year-old Thomas Edmondston.

HOW TO GET THERE

By sea Car and passenger ferries travel from Toft on the Shetland Mainland to Ulsta in the southwest of Yell. Unst is accessible by ferry from Gutcher on Yell to Belmont on Unst. For more information contact Unst Ferries Booking

Above *Dating from 1672, the Old Haa at Burravoe on Yell opened as a museum in 1984. It houses the Bobby Tulloch Collection - until his death in 1996, Tulloch was Yell's most famous naturalist, photographer and artist.*

MUCKLE FLUGGA

This small rocky island, just north of Unst, was originally called North Unst, but in 1964 its name was changed to Muckle Flugga, which comes from the old Norse *mikla flugey*, which means 'steep-sided island'. Built on this jagged rock that is the most northerly island in the British Isles, the Muckle Flugga Lighthouse first beamed out its light on 11 October 1854. Thomas Stevenson, one of the two engineers who worked on the lighthouse project, was the father of Robert Louis Stevenson, author of *Treasure Island*, which many believe was influenced by the family's time on Unst.

Above *Surely a candidate for the Turner Prize, this crazy red and yellow bus shelter, probably the most northerly in the UK, near Baltasound on Unst is full of 70s kitsch including computer, microwave and even a visitors' book.*

Information: Voicebank (tel. 01595 743972), Bookings (tel. 01957 722259) or visit www.shetland.gov.uk

ORDNANCE SURVEY MAPS
Landranger 1:50,000 series Nos. 1 & 2

TOURIST INFORMATION
Nearest office: Lerwick Tourist Information Centre, Market Cross, Lerwick, Shetland, ZE1 0LU (tel. 08701 999440) or visit www.visitshetland.com

WHERE TO STAY
On Yell, there is B&B accommodation and camping at Windhouse. There is B&B accommodation in the northern half of Unst, and a youth hostel near the Belmont ferry in the south, which also offers camping and bike hire. For more information contact Lerwick Tourist Information Centre (see above).

ISLAND WALKS
Yell With its remote and wild interior Yell is a paradise for hillwalkers. There is also good coastal walking along the east coast and up the Erisdale valley to the Catalina Memorial.
Unst There are Ranger Guided Walks in

both the Keen of Hamar or Hermaness National Nature Reserves: to book, contact Shetland Amenity Trust tel 01595 694688. Drive from Haroldwick across to Burrafirth and then walk up the Hermaness peninsula, where you can see the spectacular cliff scenery and the largest colony of puffins on the island. Just off shore at the most northerly tip of the British Isles is the island of Muckle Flugga and its famous lighthouse.

Above *This row of cottages overlooking the beach at The Taing near Norwick on Unst are some of the most northerly dwellings in the British Isles. Norwick is also the northern destination of the National Cycle Network Route 1 which begins at Dover.*
Below *The sheltered harbour at Cullivoe on the northeast coast of Yell has fine views across Bluemull Sound to the island of Unst. A few miles to the west where the road ends at Gloup is a memorial to the 58 fishermen who lost their lives in a storm in 1881*

FETLAR

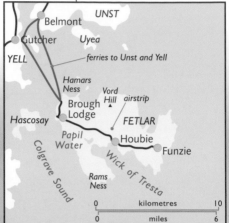

Fetlar is the smallest of the North Isles, with an area of 16 square miles, and a population of less than 100 people. It is known as the Garden of Shetland because it is very fertile – its name comes from the Norse meaning 'fat land'. There is good grazing for domestic sheep and cattle, and a rich variety of plant and birdlife. This is an island for walkers – there is no public transport on Fetlar, and no petrol or diesel fuel available.

Above *Sadly now uninhabited, Brough Lodge was built in 1820 by Sir Arthur Nicolson. Behind the house an astronomical observatory, now in a similarly ruinous state, was built by Sir Arthur on the site of an Iron Age broch.*

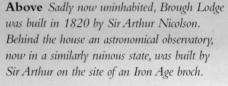

Below *Close to Tresta Beach in the south of the island, Papil Water is a favourite spot for trout fishing. In the east of Fetlar, the area around the Loch of Funzie is an important RSPB reserve and one of the last strongholds in the UK of the rare red-necked phalarope.*

HISTORY

Fetlar has been occupied from the Stone Age onwards, testified by Finnigert Dyke, a large Bronze Age stone wall. A Bronze Age standing stone at Leagarth is known as the Ripple Stone and, north of Skutes Water, is a ring of stones known as the Hjaltadans. It is also said that Gruting on Fetlar was the site of the first Norse landing in the west. The Giant's Grave, near Aith, may be the site of a Viking boat burial.

In the 1820s, the British diplomat Arthur Nicolson built Brough Lodge (now a ruin) and a tower, which he used as an observatory.. Here, a house called Leagarth was built in 1900 by Sir William Watson Cheyne who, with Lord Lister, pioneered antiseptic surgery.

NATURAL HISTORY

The coastline of Fetlar is famous for its natural arches – more per mile than anywhere else in Shetland. There are 300 species of flowering plants on the island, including the rare water sedge, which can be seen at Papil Water. East Whalsay is mainly bogland. There are otter and seal along the coast, while inland, brown trout abound in the freshwater lochs.

Fetlar is home to the rare red-necked phalarope, which breeds in the loch near Funzie, along with red-throated diver and whimbrel. The whole island is good for birdwatching and, in 1967, a pair of breeding snowy owls were discovered in the RSPB Reserve around Vord Hill. The north cliffs of the reserve are home to large colonies of breeding seabirds, including auk, gull and shag, and seal can be seen on the beaches in late autumn.

HOW TO GET THERE

There are regular car and passenger ferries between Oddsta in the northwest of the island and Gutcher on Yell and Belmont on Unst.

ORDNANCE SURVEY MAPS

Landranger 1:50,000 series No. 1

TOURIST INFORMATION

Nearest offices: Lerwick Tourist Information Centre, Market Cross, Lerwick, Shetland, ZE1 0LU (tel. 08701 999440) or visit www.visitshetland.com

WHERE TO STAY

There is limited accommodation on Fetlar, but there are some bed and breakfast establishments and a campsite.

ISLAND WALKS

With virtually no traffic, Fetlar is a walker's paradise. However, the North Fetlar RSPB Reserve has restricted access during the summer months. For more details contact the warden (tel. 01957 733246)

WHALSAY

Whalsay lies of the east coast of Shetland's Mainland. It is one of the more densely populated islands and home to the most northerly golf course in the United Kingdom. Whalsay was called 'Heals-oy' (island of whales) by the Vikings, but is referred to as the Bonnie Isle by the locals. Fishing has been the mainstay of the island for generations, with the harbour at Symbister housing its main fishing community.

HISTORY

Two neolithic houses, Yoxie and the Beenie Hoose, show that Whalsay has been occupied for at least 4,000 years. There is also an Iron Age blockhouse overlooking the Loch of Huxter.

The Bruce family took over most of Whalsay in the 1500s, dominating it for over 300 years. They went bankrupt in the 1830s, when Robert Bruce of Symbister built the New Haa (Symbister House). It cost more than £30,000 because the

granite blocks were brought from a quarry three miles away.

By the 1600s, German merchants had made their mark on Whalsay, sailing from Hamburg, Bremen and Lubeck to trade seeds, cloth, salt and iron tools for the island's dried and salted fish.

The fame of the island's fisherfolk sometimes got them into trouble. In the Napoleonic era, Royal Navy ships would intercept fishing boats from Whalsay and press the experienced sailors into service. The islanders were also hit by natural disasters such as the two fierce storms of 1832 and 1840 that claimed many lives. Today, the fishing community still thrives.

The island was home to the Scots poet Christopher Grieve, known as Hugh MacDiarmid, throughout the 1930s. In 1942, he was called up for war work and never returned to the island.

NATURAL HISTORY

The tidal sounds and the rocks offshore around the island are the best place to see the sea mammals that the

Left *Now a museum, the Pier House at Symbister is a beautifully restored Hanseatic booth which was used by German traders for several centuries until 1707.*

island is famous for – porpoise and dolphin, minke whale and orca. The island is also good for birdwatchers as it is visited by many migrants, including wheatear, meadow pipit, snow bunting and buff-breasted sandpiper.

HOW TO GET THERE

By sea A regular passenger and vehicle ferry operates from Laxo (and occasionally from Vidlin) on the Shetland Mainland. For more details visit: www.shetland.gov.uk

ORDNANCE SURVEY MAP

Landranger 1:50,000 series No.2

TOURIST INFORMATION

See facing page.

WHERE TO STAY

There is limited accommodation on Whalsay. For more details contact Lerwick Tourist Information Centre (see above).

ISLAND WALKS

Although Whalsay's west coast is fairly developed, there is easy coastal walking together with great cliff scenery and fine views across to the Out Skerries.

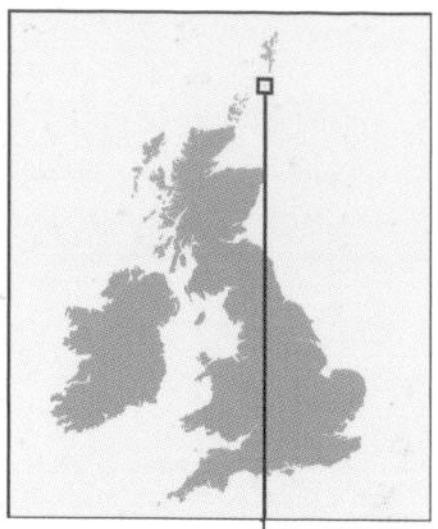

FAIR ISLE

Owned by the National Trust for Scotland since 1955 and located midway between the Shetlands and Orkneys, Fair Isle is renowned for its birdlife, wildflowers and natural beauty. While still retaining traditional crofting methods for farming, the islanders have also embraced modern technology for their energy requirements. The population of around 70 on this isolated island should be congratulated for their efforts at self-sufficiency.

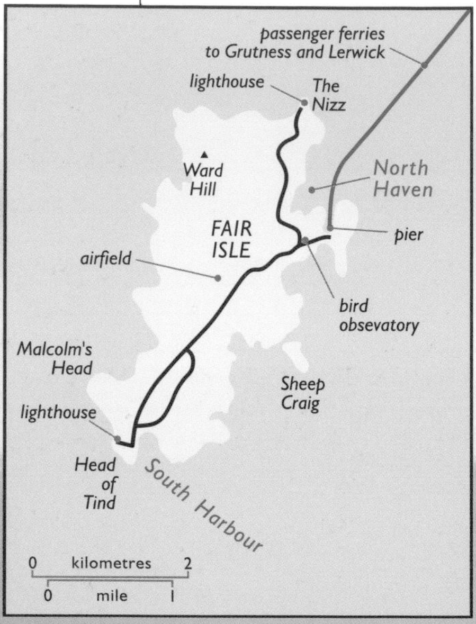

[Map labels:] passenger ferries to Grutness and Lerwick; lighthouse; The Nizz; Ward Hill; North Haven; FAIR ISLE; pier; airfield; bird observatory; Malcolm's Head; lighthouse; Sheep Craig; Head of Tind; South Harbour; kilometres 0 — 2; mile 0 — 1

HISTORY

Archæological remains, including unusual mounds of burnt stones, field systems and outlines of Iron Age houses, found on Fair Isle show that the island has probably been occupied for at least 5,000 years. In the middle of the island the Feely Dyke, a turf and stone wall, may also date from early times, separating the northern grazing land from the southern cultivated half. The remains of an Iron Age fort can still be seen at Landberg in the northeast of the island.

During Viking times, the island was probably used as a lookout post, with signalling beacons being lit in times of emergency from the summits of Malcolm's Head and Ward Hill.

For centuries, the rocky coastline of Fair Isle has been the final graveyard for many a ship. The most famous was the wreck of the Spanish galleon *El Gran Grifon*, which ran aground in 1588 after escaping north during the débâcle of the Spanish Armada. Around 250 crew struggled ashore, vastly outnumbering the island's population until they were shipped out to the Shetland mainland.

FOULA

With its enormous red sandstone sea cliffs reaching well over 1,000ft in height on its west coast, the outline of Foula is clearly visible from the Shetland Mainland. Foula is privately owned, and its small population of around 40 people derive their income from fishing, sheep raising and knitwear. Strangely, the islanders still use the old Julian calendar with Christmas Day being celebrated on 6 January! The island is also famed for its enormous seabird breeding population, including thousands of pairs of the fearsome great skua or 'bonxies' as they are popularly known. Foula can be reached from the Shetland Mainland by air from Tingwall Airport (tel. 01595 840246) or by sea from Walls (tel. 07881 823723).

Above and right *The original airstrip on Fair Isle was built during World War II to service a radar station that was built on Ward Hill. Today, the island is served by regular flights on Islander aircraft from Tingwall and Sumburgh on the Shetland Mainland.*

Together with fishing and the world-famous Fair Isle knitting, crofting has been an important part of the island economy for centuries. Even today, the southern fertile half of the island is farmed in this traditional way.

Ownership of Fair Isle changed hands several times between the 17th and 20th centuries, its population rapidy depleting, until the island was bought by the famous ornithologist George Waterston after WWII. In recognition of the island's important bird population - nearly 350 species have been recorded - Waterston also founded the world famous Fair Isle Bird Observatory in 1948. In 1955, the National Trust for Scotland purchased the island and since then, the island's population, local industries

Below *The strenuous climb to to the summit of Malcolm's Head in the southwest of the island is rewarded with this spectacular view of Fair Isle. Once a Viking signalling station, the summit of Malcolm's Head still has the concrete remains of a World War II lookout post. In the foreground is the island's west coast with its numerous stacks and geos while on the east coast, in the distance, is the dominant geological feature of Sheep Rock with its sheer cliffs and natural arches.*

and communications with the outside world have been rejuvenated. With the building of a 60KW wind turbine in 1982, Fair Isle became the first European community to be powered with electricity from a commercially-operated wind energy scheme. Since then, a more powerful 100KW turbine has been built.

Natural History

Fair Isle is renowned for both its birdlife and wildflowers. Opened in 1948, the world-famous Bird Observatory is in the frontline of scientific research on bird migration and seabird breeding colonies. Over 350 different species of bird have been observed on the island, including many rarities that are seen during the spring and autumn migratory periods. During the breeding season in spring and summer thousands of seabirds, including kittiwake, fulmar, razorbill, gannet, shag and puffin,

jostle for space along the island's rugged coastline. The inland moorland is the territory of the fearsome great skua (or 'bonxie') and tern, who will both defend their nests with almost suicidal attacks on anyone who dares to approach. Hard hats and walking sticks are *de rigeur* forms of defence against these fearsome creatures when walking on the hillsides during their breeding season. For full details of the island's birdlife contact the Fair Isle Lodge and Bird Observatory (tel. 01595 760258) or visit their website: www.fairislebirdobs.co.uk

The seas around Fair Isle are also home to a wide variety of marine life, including common and grey seal, porpoise, dolphin, minke and killer whale.

Thanks to the traditional crofting methods still used on the island, Fair Isle is a also paradise for the botanist, with over 250 species of flowering plant having been recorded. Early summer is probably the best time to visit the island, when there are vivid displays of many types of orchid, bog asphodel, squill, sea

pink and sea campion. The upland areas of the island also support many rare alpine species.

How to Get There

By air Flights from Tingwall Airport to Fair Isle operate on Monday, Wednesday, Friday and Saturday during the summer. The frequency is less during the winter. A single flight also operates on Saturday during the summer from Sumburgh Airport to Fair Isle. For more details contact Directflight Ltd at Tingwall Airport (tel. 01595 840246) or visit www.shetland.gov.uk

By sea The *Good Shepherd IV* ferry service operates from either Grutness or Lerwick to Fair Isle on Tuesday, Thursday and Saturday during the summer months. A less frequent service operates during the winter. For more details tel. 01595 760222 or visit www.shetland.gov.uk

Ordnance Survey Map

Landranger 1:50,000 series No. 4

TOURIST INFORMATION
Nearest offices: Lerwick Tourist
Information Centre, Market Cross,
Lerwick, Shetland, ZE1 0LU
(tel. 08701 999440) or visit
www.visitshetland.com
or Sumburgh Tourist Information
Centre, Wilsness Terminal, Sumburgh,
Shetland, ZE3 9JP (tel. 08701 999440).

WHERE TO STAY
Limited accommodation on Fair Isle
includes the Fair Isle Lodge and Bird
Observatory (tel. 01595 760258), several
other full board establishments and a self-
catering cottage. For more details visit
the following websites: www.fairisle.org.uk
or www.fairislebirdobs.co.uk
or www.scotland.org.uk

ISLAND WALKS
With only one single track road and
hardly any traffic, Fair Isle is a walker's paradise. Even the daytripper can
visit most of the island on foot during a
visit. Highlights include the strenuous
climb to the summits of Malcolm's Head
and Ward Hill, from where there are
panoramic views of the island, Sheep
Rock with its natural arches on the east
coast and the spectacular and rugged west
coast. Free guided walks are also
organised by the rangers at the Bird
Observatory (tel. 01595 760258).

Above *Traditional crofting methods are still
the order of the day in the cultivable southern
half of Fair Isle. Self-sufficiency is a way of
life, with vegetables being grown in modern
polytunnels.*

Below *In the northeast, North and South
Haven are separated by a narrow isthmus. The
ferry service from Grutness on the Shetland
Mainland to the pier at North Haven is
provided by the sturdy Good Shepherd IV,
which is hauled on to a slipway for safety
during bad weather.*

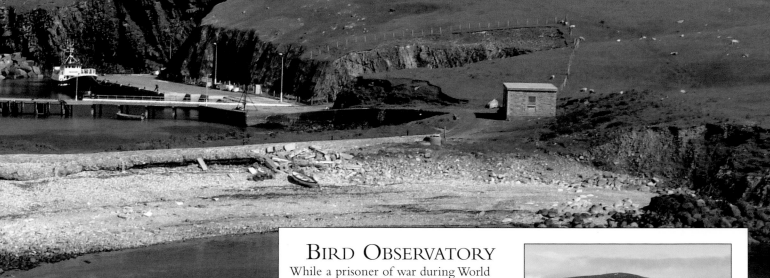

BIRD OBSERVATORY
While a prisoner of war during World
War II the famous ornithologist George
Waterston made plans to set up a bird
observatory on Fair Isle. His dream came
true in 1948 and since then the
observatory has become world-famous for
its scientific study of migratory birds. It
has recently been modernised and
provides accommodation and facilities for
visitors who are attracted to the island's
wide-ranging species of bird that can be
seen during the spring and autumn
migratory periods. Fair Isle is also an
important seabird breeding site and
during the breeding season the nearby
cliffs are alive with the comings and
gopings of thousands of fulmar, gannet,
razorbill, guillemot, shag and puffin.

Fair Isle Lodge and Bird Observatory

A Heligoland trap used for ringing birds

Below *Deserted and unspoilt beaches are a feature of the Outer Hebridean islands. This beach, close to the remote and tiny settlement of Gleann Dail on the south coast of South Uist, overlooks the narrow Sound of Eriskay to the island of that name. Eriskay is world-famous as the location for the real-life* Whisky Galore *when the SS Politician ran aground here with its precious cargo of malt whisky in 1941.*

Outer Hebrides

Lewis and Harris

North Uist

Benbecula

South Uist

Barra and Vatersay

St Kilda

LEWIS AND HARRIS

Although Lewis and Harris, one of the last great wildernesses in the British Isles and by far the largest of the Scottish islands, are one island, they are effectively separated by a large mountain chain. Both geographically and historically each has its own unique identity, but one thing they do have in common is religion – this is one of the last outposts of the UK to adhere strictly to Sunday observance.

[Map: Isle of Lewis and Harris showing kilometres/miles scale, locations including lighthouse, Butt of Lewis, Ness, Fivepenny Borve, North Tolsta, Barvas, Gress, Tolsta Head, Carloway, Dun Carloway, Broad Bay, Tunga, Stornoway Airport, Tiumpan Head, GREAT BERNERA, Breasclete, ISLE, Eye Peninsula, Loch Roag, Callanish, OF, Stornoway, Garrabost, Uig, LEWIS, Mealisval, Tohaval, Loch Erisort, vehicle ferry to Ullapool, SCARP, Loch Langavat, Seaforth Island, Beinn Mhór, Husinish Bay, Clisham, Loch Seaforth, Sound of Shiant, Ben Raah, Beinn Dhubh, Tarbert, Shiant Islands, TARANSAY, Toe Head, bridge, SCALPAY, Chaipaval, HARRIS, passenger ferry to North Uist, Leverburgh, vehicle ferry to Lochmaddy, Pabbay, vehicle ferry to Uig, Berneray, Rodel]

HISTORY

Probably visited by the nomadic hunters of the Middle Stone Age around 7,000 years ago, Lewis and Harris have been permanently inhabited for over 5,000 years. In particular, Lewis has many prehistoric sites, including Neolithic burial cairns, standing stones and stone circles, of which the most famous is the complex found at Callanish on the west coast of Lewis (see below). Once covered in peat, many more of these sites are still being uncovered by archæologists. Many Bronze Age artefacts, such as pottery and weapons, have also been uncovered on the island and there is no doubt that more will be found in the future, buried beneath the peat. The Iron Age is well represented by several coastal promontory forts and duns, of which the most famous is the superbly preserved example at Carloway, north of Callanish on the west coast of Lewis.

By the late 8th century AD, the region had increasingly come under attack from Viking raiders. Some of these Norsemen

Above *Stornoway, the capital and administrative centre of Lewis and Harris, is the only large town in the Western Isles. A major fishing port during the 19th century, the town and its immediate surrounding area has seen its population decline in recent years to a low of around 8,000 - 30% of the total population of the Western Isles! Lews Castle, seen in the distance, was built in the mid-19th century by Sir James Matheson from a fortune made in the Far Eastern opium trade. Now awaiting restoration, it served as a naval hospital during World War II and then as a Technical College until 1989. Its 600 acres of grounds of mixed woodland are open to the public.*

Below *As famous as, and probably older than Stonehenge, this megalithic stone circle, with its 15½ft-high central stone and four avenues of standing stones is located at Callanish on the west coast of Lewis overlooking Loch Roag. Composed of locally quarried Lewisian gneiss, they were erected around 5,000 years ago and many theories, including an ancient temple or a calendar system linked to astronomical movements, have been put forward for their existence. During archæological excavations in 1857, large amounts of peat were removed from the site. This not only revealed the full height of the stones, but also a burial cairn within the central circle. This historic site, in the care of Historic Scotland, with its own car park and visitor centre, is signposted from the A858.*

Right *Dun Carloway is located in a well-defended position on a commanding hilltop overlooking Loch Carloway on the west coast of Lewis. A masterpiece of drystone walling, this Iron Age broch is one of the finest examples of its kind in the Western Isles. Built around 2,000 years ago, it is now peeled open to reveal its double-skinned walls, staircases and floor levels.*

eventually settled on Lewis and Harris and various artefacts from this period, such as jewellery and a beautifully carved ivory chess set, have been discovered on the island.

Norse rule came to an end in 1263 after the Norwegians were defeated by Alexander III of Scotland at the Battle of Largs. Three years later, the Treaty of Perth was signed and the Western Isles, including Lewis and Harris, were sold to the kingdom of Scotland for 4,000 marks. In 1335, Lewis and Harris became part of the Hebridean kingdom ruled by the MacDonalds or Lords of the Isles. In 1493, however, following years of unrest in the Western Isles, King James IV of Scotland took over the title of Lord of the Isles for the crown and since then the heir to the throne has held the title.

On the forfeiture of the Lordship of the Isles, Lewis and Harris was granted to the MacLeods who held on to it until 1610, when Lewis was granted to the MacKenzies of Kintail. Headed by the Earls of Seaforth, the earlier MacKenzies developed Lewis into a massive sporting estate and treated their tenants pretty badly. In 1844, Lewis was sold to Sir James Matheson, who had made his fortune from the Far Eastern opium trade. Matheson pumped large amounts of his personal wealth into improving the island, building roads and generally improving the infrastructure and, by the mid-19th century, Stornoway had become one of the most important fishing ports in Europe. In 1918, Matheson's descendants sold Lewis to the

soap magnate Lord Leverhulme, who had grandiose schemes to further develop the island.

The MacLeods continued to hold on to Harris until 1834 when it was sold to the Earl of Dunmore, who then went on to sell North Harris to Sir Edward Scott. It was around this time that the Harris Tweed industry grew from obscurity to become a world-famous product – there are still weavers today on Harris that produce this heavy cloth in the traditional manner. Unable to pursue his development of Lewis, Lord Leverhulme first gave away much of his land to the

islanders before concentrating his efforts on Harris which he had bought in 1919. His efforts to develop Leverburgh (originally named Obbe but renamed after him) into a major fishing port never fully materialised, due to his death in 1925.

With the great benefactor gone, Lewis and Harris was split into estates which were sold off in piecemeal fashion. The consequence of the loss of Lord Leverhulme's investment in the island was large scale unemployment and subsequent mass emigration to Canada for many young families.

Today, traditional industries such as

THE SUMMER ISLES

The Summer Isles are a group of small islands located about 12 miles northwest of the mainland ferry port of Ullapool. The nearest point on the mainland is the village of Achiltibuie. Boat trips to the islands operate from Ullapool on the *Summer Queen* (tel. 01854 612472; website: www.summerqueen.co.uk) or from Achiltibuie on the *MV Hectoria* (website: www.summer-isles-cruises.co.uk). The largest and only inhabited of the islands, Tanera Mor, is famous for its otters, seals and seabirds. In addition to a fish farm, there is also a post office on the island, which issues its own unique Summer Isles stamps. A self-catering cottage is available to rent. For more details about Tanera Mor, visit the island's website: www.summer-isles.com

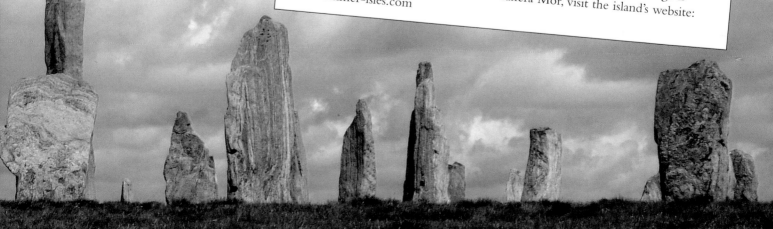

weaving, crofting and fishing, along with a growing tourist industry, are the main economic activities on the island.

Religion is still a very important aspect of many islanders' lives. Lewis and Harris is predominantly Protestant, while the southern isles of the Outer Hebrides are mainly Roman Catholic. Calvinism dominates the Protestant church on Lewis and Harris and visitors to the island should note that the Sabbath is still strictly observed - ie everything shuts down for the day!

NATURAL HISTORY

Although Lewis and Harris are one island, they are effectively separated by a large mountain chain, rising to a height of over 2,000ft, that runs from Loch Resort on the west coast to Loch Seaforth on the east coast.

Lewis

In the north, Lewis was once covered in scrubby deciduous woodland but, due to both man's activities and climate change over the last 5,000 years, is now a vast region of large freshwater lochs and

Right *Sheltered by the small island of Eilean Iuyard at the entrance to Loch Sealg, the remote fishing village of Lemreway is reached at the end of a long and winding road in the Lochs region of eastern Lewis. To make way for more profitable large-scale sheep farming, Lemreway's population fell victim to the infamous 'clearances' of the 19th century, dropping from 179 in 1841 to a depressing 18 in 1851.*

treeless peat moors. Lewis' coastline can also be split into two physically different types - the northern half consists mainly of sea cliffs that rise to a height of over 200ft in places, while the rugged southern half is punctuated by many inlets and sea lochs.

Although much larger than Harris, Lewis (roughly 38 miles long x 33 miles wide) does not have such a wide range of natural habitats as its neighbour. Most of its interior is a patchwork of desolate and uninhabited peat bog, while its coastline has much less of the sand dunes and machair found further south. In recent years, the escape of animals from mink farms on Lewis has sadly brought about

the demise of many species of bird and small mammal. Despite all this, there are still many areas of natural interest to be found on Lewis.

One of the highlights is the extensive grounds of Lews Castle in Stornoway, where the mixed woodland, planted by Sir James Matheson in the 19th century, includes rare foreign species of tree and is home to many woodland birds usually only found on the mainland.

Other areas of interest in the eastern half of Lewis include Loch Branahuie, between Stornoway and the Eye

Below *On the west coast of Lewis, the magnificent sweep of Uig Bay is dominated by the island's mountainous interior with its highest peak, Mealisval, rising to a height of 1,883ft. In 1831, a magnificent 78-piece Viking chess set was discovered in sand dunes near Uig Sands. Dating from the 12th century, the set is beautifully carved from walrus tusk ivory. Although the originals are now in mainland museums, copies of the chess set can be seen in Stornoway Museum.*

Peninsula. This is a good place to see many species of seaduck, including widgeon, red-breasted merganser and tufted duck and, during autumn and winter, the black-throated and great northern diver. Tiumpan Head, at the eastern point of the Eye Peninsula, is a popular spot for whale and dolphin spotting.

One of the few large stretches of saltmarsh, sand dunes, machair and beach to be found on Lewis are found immediately to the north of Stornoway Airport.

During summer this area, which includes Melbost Sands and the estuary of the River Laxdale, is home to breeding tern, ringed plover, shelduck and oystercatcher while, in winter, it is visited by hundreds of wader, geese and duck. Large numbers of eider duck also overwinter further up the coast at Coll and Gress. The vast area of moorland to the northwest of Gress is also the breeding ground for the great and arctic skua. North of Tolsta Head, Garry Sands is famous for the three sea

stacks that are exposed here at low tide.

At the far north of Lewis is an area known as Ness, which ends at the Butt of Lewis lighthouse. Lashed by Atlantic storms during the winter, the Butt is a good place for dolphin and whale watching and to witness the spectacular diving of gannet as they head home to their colony on the island of Sula Sgeir, a National Nature Reserve over 40 miles to the north.

Inland from the Port of Ness Loch Stiapavat, one of the finest locations for wildfowl on Lewis, is

GREAT BERNERA

Linked to the west coast of Lewis by a concrete bridge (via the B8059) since 1953, the small island of Great Bernera (6 miles x 3 miles) is owned by Count Robin de la Lanne Mirrlees (also known as Prince of Incoronata) – an anti-monarchist and Buddhist. Based at Kirkibost on the shores of Loch Roag, Great Bernera has been an important centre of the local lobster fishing industry for around 200 years. The island is rich in archæological remains including several standing stones, located immediately after crossing the bridge (right), and the remains of an Iron Age village that were recently discovered at Bostadh. Breacleit, the main community in the centre of the island, houses a small museum which includes information on the famous Bernera Riots of 1874, when the island's crofters refused to be moved from their lands. For more details on Great Bernera visit the island's website: www.uigandbernera.com

Above *Room with a view - overlooking the now-uninhabited island of Scarp (on left) from the north side of the isthmus at Husinish on the northwest coast of Harris. It was from Scarp in 1934 that the German rocket engineer Gerhard Zucker (1900-1985) tried to demonstrate, unsuccessfully, that rockets could be used to deliver mail to remote locations.*

Below *With its white sandy beach and crystal clear water, Husinish Bay is as close as you can get to Paradise on a fine day on Harris. From left to right on the horizon are the island of Taransay, Toe Head peninsula on South Harris and the island of Pabbay.*

home to mallard, teal, widgeon, tufted duck, whooper swan and black-headed gull. Much of this northerly region is still farmed using traditional crofting methods and, as such, provides the necessary ground cover in summer for the elusive corncrake.

South from Ness and bounded by the A857 to the west and by seacliffs to the east, is a vast and remote region of freshwater lochs, home to brown trout and wildfowl, and peat bogs, home to many rare plant species including bog asphodel, bog cotton, sundew and deer grass. Dotted with the occasional ruins of long-deserted townships this region, difficult and dangerous to traverse on

foot, is one of the last great untouched wildernesses in the UK.

The RSPB nature reserve at Loch na Muilne, near Arnol on the west coast, is an important breeding site for the red-necked phalarope. In the spring and summer it is home to many breeding waders and, during winter, provides shelter for whooper swan, snipe and geese. Further south on the far west coast, the stunningly beautiful sandy beaches at Valtos, Uig and Carnish, with their hinterland of dunes, machair and reedbeds, contrast with the mountain range that looms over them from the south. During summer, the machair along this coastline is a colourful riot of wildflowers including many rare orchids.

SCALPAY

Scalpay lost its unique island status when a new road bridge, linking it to South Harris, was opened by Tony Blair in 1997. Avoiding the infamous 'clearances' of the late 18th and early 19th centuries, the island with its sheltered north and south harbours has for centuries been an important base for the local shellfish industry. With a population of over 300 'Scalpachs', this small island (3 miles x 2 miles) has a fairly high population density compared to its neighbours. For walkers, a 'Heritage Trail' wends its way from the main west coast settlement around the North Harbour (right), through a patchwork of moorland and freshwater lochs and skirting the island's highest point of Beinn Scoravick (341ft), to Eilean Glas Lighthouse. The lighthouse was the first of its kind in the Western Isles when it became operational in 1789. The present building, designed by Robert Stevenson, dates from 1824. Disused since 1987, the foghorn was installed in 1907. For more details about Scalpay visit the island's website: www.scalpay.com

Even further south, the steep seacliffs near Mangarista, which reach a height of over 200ft, are home to many rare cliff plants such as roseroot.

Bounded by the sea to the east and south, Loch Erisort to the north and Loch Seaforth to the west, Park is a vast area of uninhabited upland. With many peaks reaching well over 1,000ft (Beinn Mhor is the highest at 1,874ft) and hundreds of small freshwater lochs, this region is so named as it became the deer park for the island's Laird during the 17th century. With its large herds of red deer, Park is still used during the season for deer stalking.

Harris

Shaped by the action of glaciers during the last Ice Age, the mountains of North Harris are the natural, physical boundary between Lewis and Harris. From Loch Resort in the west to Loch Seaforth in the east this formidable barrier, with the mountain of Clisham (2,622ft) being the highest, is only traversed by one road – the A859. With their many glacial features, these remote mountains are also home to the golden eagle. Reached via a tortuous and winding coastal road from Ardhasig Bridge the crystal clear water, white shell-sand beach and machair hinterland at Husinish is one of the most beautiful and remote locations on North Harris. A riot of wildflowers in late spring and early summer, there are good

views from the jetty at Husinish to the uninhabited island of Scarp.

The west coast of South Harris is famed not only for its vast areas of sand which are exposed at low tide, but also for its enormous tracts of high sand dunes. A favourite haunt of wading birds and wildfowl, the sand flats at Luskentyre are rich in marine life while, further south around Northton, the estuarine salt marshes and coastal machair, a riot of colourful wildflowers in early summer, attract many species of breeding bird including skylark, lapwing, redshank and, during winter, hundreds of geese.

Finally, in complete contrast, the glaciated landscape of eastern Harris with its myriad of freshwater lochs and thousands of huge rocks (known as

Above *Located at Bunavoneadar, northwest of Tarbert on North Harris, this solitary brick chimney is all that remains of a whaling station originally built by a Norwegian company to process whale carcasses in the early 20th century. The station was bought by Lord Leverhulme, the owner of Harris from 1919, in the early 1920s and continued in use until around 1930. Dominated by the North Harris mountains, the chimney is now classed as a Scheduled Industrial Monument by Historic Scotland.*

erratics) left stranded by melting glaciers around 10,000 years ago, gives the visitor the strange feeling that they have landed on another planet! Amazingly, through their own ingenuity, humans have survived in this seemingly inhospitable

region for hundreds of years. Seals are a common sight in the many small inlets and on islets along this rugged eastern coastline.

For further information about wildlife throughout the Outer Hebrides visit Wildlife Hebrides website: www.wildlifehebrides.com

HOW TO GET THERE
By sea Caledonian MacBrayne operate a regular vehicle and passenger ferry between Ullapool on the mainland and Stornoway (Lewis). They also operate regular vehicle and passenger ferries between Uig (Isle of Skye) and Tarbert (Harris) and between Berneray (North Uist) and Leverburgh (Harris).

For more details on all of these services contact the CalMac reservations office (tel. 08705 650000) or visit their website: www.calmac.co.uk
By air British Airways operate regular flights from Glasgow, Edinburgh and Inverness to Stornoway. For more details contact BA (tel. 0870 850 9850) or visit their website: www.ba.com

Highland Airways operate flights from Benbecula and Inverness to Stornoway. For more details contact Highland Airways (tel. 0845 450 2245) or visit their website: www.hial.co.uk

Eastern Airways operate flights from Aberdeen to Stornoway. For more details contact Eastern Airways (tel. 08703 669 100) or visit their website: www.easternairways.com

ORDNANCE SURVEY MAPS
Landranger 1:50,000 series Nos. 8, 13, 14 & 18

TOURIST INFORMATION
Lewis: Stornoway Tourist Information Centre, 26 Cromwell Street, Stornoway, Isle of Lewis HS1 2DD (tel. 01851 703088).
Harris: Tarbert Tourist Information Centre, Pier Road, Tarbert, Isle of Harris HS3 3DJ (tel. 01859 502011).
For either of the above also visit their website: wwwvisithebrides.com

WHERE TO STAY
There is a wide range of accommodation on Lewis and Harris. For more details contact either Stornoway or Tarbert information centres (see above) or visit their website: www.visithebrides.com

ISLAND WALKS
Roughly 61 miles at its longest and 33 miles at its widest, Lewis and Harris, one of the last wildernesses in the UK, is ideal for a walking holiday and offers an enormous choice of walking opportunities,

TARANSAY

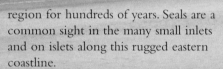

Located just over one mile from the northwest coast of South Harris, the twin-peaked island of Taransay with its white sandy beaches shot to fame in 2000 when it was chosen as the location for the BBC TV series *Castaway*. It has also been used more recently as one of the locations for the film *The Rocket Post*. Privately-owned and run as a working sheep farm, Taransay is also the site of two early Christian chapels, one dedicated to St Taran and the other to St Keith. Two self-catering units are available to rent on this otherwise uninhabited island. For more details visit the island's website: www.visit-taransay.com

Above *Exposed at low tide, the vast expanse of Luskentyre Sands in South Harris is interlaced by several small rivers, which are fed from inland lochs such as Loch Carran. In the distance, the mountains of North Harris form a natural barrier between Lewis and Harris. A paradise for hillwalkers, the four highest peaks in this range (from west to east) are Tirga Mor (2,228ft), Oreval (2,165ft), Uisgnaval (2,392ft) and Clisham (2,622ft).*

ranging from miles of sandy beaches, dunes and machair to thousands of acres of remote moorland, hills and mountains. The Heritage Trail is a 10-mile coastal walk along the remote east coast of Lewis from Garry Beach to Ness.

The island's two tourist information centres have booklets and leaflets for many waymarked walks and can supply experienced guides for organised walks. Hostels and traditional blackhouse cottages in remote locations provide suitable accommodation for walkers. For full details of many fascinating walks on Lewis and Harris, visit the first website dedicated to walking in the Outer Hebrides: www.walkhebrides.com

Above *Located in the village of Rodel at the southernmost tip of the Isle of Harris, St Clement's Church is one of the most spectacular in the Western Isles. Built on the site of an earlier church by Alastair (Hunchback) MacLeod of Harris in the early 16th century, St Clement's became the burial place for the MacLeods of Harris and contains many fine tombs and carved gravestones. It is now a place of pilgrimage for MacLeods from around the world.*

NORTH UIST

Apart from a few peaks, much of North Uist is covered in moorland, bog and hundreds of small lochs. The island, rich in early archaeological remains, was ruled for 500 years by the MacDonalds of Sleat until 1855, when it was sold to Sir John Powlett Ord. While over 2,000 islanders had been ruthlessly evicted by their previous laird, those that were left fared little better. Today most of North Uist, with its much reduced population of around 1,600, is owned by the North Uist Estate Trust.

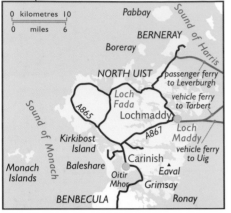

HISTORY

North Uist is particularly rich in early archæological remains that show that the island has been inhabited for over 6,000 years. The Neolithic period is represented by the well-preserved Barpa Langass burial mound and the nearby Pobull Fhinn stone circle, which can be seen close to the A867 just over two miles northeast of Clachan. They are probably two of the oldest surviving stone structures in the Outer Hebrides. There are many other chambered cairns and standing stones from this period dotted around the island.

The later Iron Age is also well represented by the remains, in the form of mounds, of the many duns (fortified houses) that were built on artificial islands

Left *Located at Carinish in the south of North Uist, Trinity Temple was once an important monastery and college of international standing founded by Beathag, daughter of Somerled. It was enlarged in the late 14th century by the wife of John, Lord of the Isles, but destroyed after the Scottish Reformation. Although restored in the 19th century, it is now in a very sad state of repair.*

in the middle of freshwater lochs. Good examples on North Uist are Dun an Sticer, near Port nan Long in the north of the island, and on Loch an Duin and Loch na Caiginn near Lochportain in the east.

Surprisingly, very little has been found on North Uist from either the period of the early Celtic Christians or the Vikings, who ruled the region from the 9th to the 13th centuries. By this time, however, North Uist had became an important place of learning through the monastery that was founded by Somerled's daughter, Beathag, at Trinity Temple near Carinish.

Following the end of Norse rule, North Uist soon came under the control of the MacDonalds of Sleat on Skye who managed, come hell or high water, to hold onto the island until 1855 when it was sold to Sir John Powlett Ord. Sadly by then, with the disintegration of the local kelp industry, potato blight and crop failures, over 2,000 starving and impoverished islanders had been ruthlessly cleared from their farms and homes by Lord MacDonald and shipped across the Atlantic to Canada.

Lochmaddy, the 'capital' of North Uist, once a haven for pirates, first came

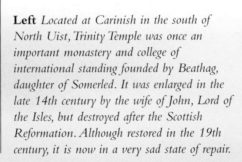

MONACH ISLANDS

A group of five small low-lying islands located in the Atlantic five miles west of North Uist, the Monach Islands were probably inhabited for over 1,000 years. Once home to a monastery and, by the late 19th century, to nearly 150 people who survived on sheep raising and fishing, the Monachs have been uninhabited since 1948 when the lighthouse was taken out of use. Now owned by North Uist Estates, the islands are designated as a National Nature Reserve and managed by Scottish Natural Heritage. The islands' machair is home to nearly 200 species of plants and grasses and is an important breeding ground for cormorant, shag, fulmar, eider duck, Arctic tern and black guillemot. In addition to its large grey seal colony, the Monachs are also an important overwintering destination for barnacle geese. For more details on the Monach Islands visit the website: www.monachs.com

to prominence in the 17th century as an important centre for the herring processing industry. Providing much employment, this peaked in the late 19th century but, by the 1930s, had ceased to exist due to overfishing. Today the village of Lochmaddy, with its hotel, museum and arts centre, is the main link with the mainland via the ferry service to Uig on the Isle of Skye.

NATURAL HISTORY

With its wide range of habitats, North Uist is particularly rich in plant, bird and marine life. The island's Atlantic west coast has one of the longest continuous stretches of sand dunes and machair in the Outer Hebrides. Here, the large RSPB Nature Reserve at Balranald, established in 1966, is famed worldwide for its amazing number of rare plant and bird species that can be seen throughout the year. Balranald's rich variety of habitats, contained within a traditional crofting landscape, provides ideal conditions for many rare wading and farmland birds such as the corncrake and corn bunting. The reserve is also an important site for many rare wildflowers, grasses and insects. Otters are regulary seen in the reserve's freshwater lochs and along the rocky shoreline, from where visitors can also often witness the antics of porpoise, dolphin and minke whale.

On the opposite side of the island, the eastern coastline is indented by a myriad of lochs of which the largest, Loch Maddy, with its islets, lagoons and coastline has been designated a Marine Special Area of Conservation. In particular, its obs - lagoons that have a constantly changing mixture of fresh and sea water, separated at low tide from the sea by rocky sills - are internationally important for their rich variety of plant and animal life. This important marine reserve is managed by Scottish Natural Heritage.

HOW TO GET THERE

By sea Caledonian MacBrayne operate a regular vehicle and passenger ferry between Uig (Isle of Skye) and Lochmaddy. They also operate a regular vehicle and passenger ferry between Leverburgh (Harris) and Berneray. North Uist is linked to the island of Bernerary (see below) by a road causeway.

For more details on all of these services contact the CalMac reservations office (tel. 08705 650000) or visit their website: www.calmac.co.uk

By air There are regular flights to Balivanich Airport on Benbecula. From here North Uist is reached via a road causeway. For more details see page 67.

ORDNANCE SURVEY MAPS

Landranger 1:50,000 series Nos. 18 & 22

TOURIST INFORMATION

Nearest office: Lochmaddy Tourist Information Centre, Pier Road, Lochmaddy, North Uist, Western Isles HS6 5AA (tel: 01876 500321) or visit website: www.visithebrides.com

WHERE TO STAY

Accommodation on North Uist ranges from several hotels and guesthouses to bed and breakfast and self-catering establishments and a hostel. For more details contact Lochmaddy Tourist Information Centre (see **TOURIST INFORMATION**).

ISLAND WALKS

There are many opportunities for walking on North Uist, including strenuous climbs to the summits of Eaval (1,138ft), virtually an island in itself, the twin peaks of Li a Tuath (860ft) and Li a Deas (922ft), all of which are in the southeast, and Beinn Aulasaraigh (712ft) in the centre of the island. Less strenuous is to walk (at low tide - check tide times first!) across the expanse of Vallay Strand to the uninhabited island of Vallay with its ruined house built in the 19th century by the wealthy mill owner and antiquary, Erskine Beveridge (1851-1920). For more details of these and other fascinating walks visit the excellent website: www.northuist.net and click on Walks. Details of self-guided walks around Lochmaddy - designated as a Marine Special Area of Conservation - are available from the Taigh Chearsabhagh arts centre in the village. Between May and August, guided walks are provided by the RSPB at their extensive Balranald Nature Reserve on the west coast. For more details contact the reserve manager (tel. 01876 560287).

Below *Located in the northwest of North Uist, Scolpaig Tower is a folly that was built on the site of an Iron Age broch in the 1830s by Dr Alexander MacLeod. The original broch, the stone of which was used to build the folly, was the scene of a murderous family feud between members of the MacDonalds during the 16th century.*

BERNERAY

Recently linked by a road causeway to North Uist and a passenger and vehicle ferry to Leverburgh on Harris, the small island of Berneray (3 miles x 2 miles) is famed not only for its three-mile long white shell sand beach but also for one of the finest stretches of machair in the Outer Hebrides. A colourful riot of wild flowers during the summer, the machair is designated as a Site of Special Scientific Interest and a Special Protection Area of international importance for birds. There is a waymarked walk that incorporates parts of the machair and the beach. For more details of this unspoilt island, visit the island's website: www.isleofberneray.com

BENBECULA

Now linked by causeways to North and South Uist, Benbecula was for centuries a stronghold of the all-powerful Clan Ranald, until falling into the hands of the infamous Colonel Gordon of Cluny in the 19th century. Following a century of decline, the building of an airfield, military base, local authority administrative centre and new schools have all contributed to a reversal in population decline and brought about brighter prospects for the islanders' future.

HISTORY

Archæological remains, such as the Neolithic standing stone at Gramasdail and the Iron Age broch at Dun Buidhe, show that Benbecula has been inhabited for at least 5,000 years. The early Celtic Christians, represented by the remains of Teampall Chaluim Chille at Balivanich, were overtaken by the Vikings, who remained in control of the region until 1266. Their descendants, the all-powerful Clan Ranald. then took control; the remains of one of their fortified houses, Borve Castle, can be seen today in the southwest of the island. The Clan Ranald hold on Benbecula only ended in 1839 when it was sold to the notorious Colonel Gordon of Cluny (see page 70).

For the next 100 years Benbecula's population went into steady decline, which was only halted at the outbreak of WWII with the building of a bridge across South Ford to South Uist and the development of Balivanich as a major RAF airfield.

Since then, a new causeway to North Uist, via Grimsay (see below), the opening of a missile testing range on South Uist, the replacement of the bridge to South Uist by a causeway, the development of Balivanich as a regional airport, military base and local authority administrative centre and the building of new educational facilities have all brought about new optimism for Benbecula's future prospects.

NATURAL HISTORY

Exposed at low tide, the vast expanses of sand of the North and South Fords are good locations to see a wide variety of wading birds while, inland, the moorland, bogs and freshwater

Below *Benbecula's remote east coast is heavily indented by many sea lochs such as Loch a' Laip near Creagastrom - a favourite haunt of otters and seals.*

GRIMSAY

Grimsay (2½ miles wide x 2½ miles long) ceased to be an island in 1960, when a road causeway was opened linking it with North Uist and Benbecula. Before then, communications between these islands depended on a ferry that could only operate at high tide, or a dangerous trip across the vast and treacherous expanse of sands, known as North Ford, which were exposed at low tide. Today, with its circular coastal road, Grimsay is an important centre for the fishing industry, which operates from the busy modern harbour at Kallin.

lochs are particularly noted as being home to breeding hen harrier, short eared owl and other birds of prey. Corncrakes are also regular visitors to the area around Balivanich. Benbecula's remote and indented eastern coastline is also a good location to spot the shy otter.

During the winter gales enormous amounts of seaweed, used for centuries to fertilise the machair and in the 18th and 19th centuries for the kelp-burning industry, are deposited along Benbecula's western beaches.

How to Get There

By sea There are no direct sea links to Benbecula. However, Caledonian MacBrayne operate regular vehicle and passenger ferries between Uig (Isle of Skye) and Lochmaddy (North Uist) and between Oban and Lochboisdale (South Uist). Benbecula is linked to both North Uist and South Uist by road causeways.

For more details on all of these services contact the CalMac reservations office (tel. 08705 650000) or visit their website: www.calmac.co.uk
By air British Airways operate regular flights from Glasgow and Barra to Benbecula. For more details contact BA (tel. 0870 850 9850) or visit their website: www.ba.com

Highland Airways also operate regular flights from Inverness and Stornoway (Lewis) to Benbecula. For more details contact Highland Airways (tel. 0845 450 2245) or visit their website: www.hial.co.uk

Ordnance Survey Map
Landranger 1:50,000 series No. 22

Tourist Information
Nearest office: Lochmaddy Tourist Information Centre, Pier Road, Lochmaddy, North Uist, Western Isles HS6 5AA (tel: 01876 500321) or visit website: www.visithebrides.com

Where to Stay
Accommodation on Benbecula ranges from several hotels and guesthouses to a few bed and breakfast and self-catering establishments. For more details contact Lochmaddy Tourist Information Centre (see above) or visit the island's website: www.isle-of-benbecula.co.uk

Island Walks
Benbecula is a flat, slightly featureless island (7½ miles wide x 7½ miles long) covered with a patchwork of hundreds of small freshwater lochs. The only high point, Ruabhal (407ft), can be reached along a lonely track that winds its way

Above *This sad, derelict house, near to Benbecula's small 19th century fishing harbour at Peter's Port, is typical of the hundreds that can be seen dotted all around the Outer Hebridean landscape.*

from the A865 around the lochs and continues on to the east coast at Scaraloid Bay and Rossinish. It was from this remote peninsula in 1746 that Flora MacDonald assisted Bonnie Prince Charlie, disguised as Betty Burke, to sail to Skye before he finally escaped to France.

The Military on Benbecula

First used as a grass airstrip in the 1930s, the airfield at Balivanich grew into a major airbase for the RAF during World War II. Between 1942 and 1945, squadrons of B17 Flying Fortresses and Wellingtons based at Balivanich flew long and lonely sorties out over the North Atlantic, seeking out German U-boats and protecting convoys. Today, the airfield is not only used for domestic civilian flights to Barra, Stornoway (Lewis), Inverness and Glasgow but also by the military in support of the missile testing range on South Uist. Becoming operational in 1958 this range, now managed by QinetiQ, with its associated facilities at Balivanich and a tracking station on St Kilda, is currently being used to evaluate the latest missile systems employed by the Eurofighter Typhoon aircraft. The RAF will eventually operate 232 of these aircraft at a total cost of around £20 billion!

SOUTH UIST

After centuries of Viking and traditional Clan rule, South Uist was sold to the notorious Colonel John Gordon of Cluny in 1838. Within a few years, half of the island's poverty-stricken and starving population had been forcibly evicted by their new landlord and shipped out to Canada. An island of geographical extremes, South Uist today is also home to one of Scotland's most important National Nature Reserves and to one of Europe's most important missile testing ranges.

HISTORY

South Uist, along with the rest of the Outer Hebrides, has been inhabited for at least 6,500 years. It was then that the first farmers cleared the islands of their native woodland to make way for grazing and crops. These early people left permanent monuments to their dead, of which the chambered cairn at Reineabhal, about three miles north of Daliburgh on South Uist, is a good example. Occupation during the Bronze Age is well represented by the stone foundations of a row of roundhouses that have been excavated at Cladh Hallan.

By the Iron Age, these stone roundhouses had developed into fairly sophisticated brochs, unique to Scotland, and remains of these can still be seen on South Uist as mounds on artificial islands in several lochs on either side of the A865 near Milton and further north on Loch Altabrug. Dating from the late Iron Age (around the 1st century AD), the remains of a wheelhouse, a type of dwelling unique to this region, can be seen at Kilpheder on the west coast.

Excavations at one of the ruined chapels at Tobha Mor, near the west coast not far from Loch Druidibeg, show that Christianity probably arrived on South Uist around the 7th century. Before it was destroyed by Viking invaders in the 9th century, it seems likely that Tobha Mor had grown into a regionally influential monastery. During Viking rule, when the Outer Hebrides became part of an enormous Norse kingdom stretching from the Isle of Man to the Orkneys, South Uist developed into an important regional trading centre and recent excavations at Bornais have uncovered goods from as far afield as the Mediterranean.

After the collapse of Norse rule in 1266, the Western Isles were ruled by their descendents, the all-powerful Clan Ranald. Set amidst the many small inland

FLORA MACDONALD

Born on a tenant farm in Milton, South Uist, Flora MacDonald took Bonnie Prince Charlie, disguised as a maidservant, from Benbecula to Skye in 1746. On arriving at Portree, the Prince left for Raasay (see pages 84–85) before catching a ship back to France. Flora never saw him again but was imprisoned in the Tower of London. On her release, she married a Skye farmer before emigrating to North America. She and her husband eventually returned to Skye where she died in 1790. A memorial (above) commemorating Flora's birth in 1722 can be seen today in the ruins of Milton township.

Below *Founded in 1958, the National Nature Reserve at Loch Druidibeg includes over 4,000 acres of land and loch that stretch across South Uist from the western coastline to the eastern mountains. Managed by Scottish Natural Heritage, the reserve contains important breeding areas for wading birds and thousands of resident greylag geese.*

Right *Typical of the landscape on South Uist's west coast, this stretch of machair, or sand dune pasture, is being seeded to provide oats and rye as fodder crops. The machair is also an important habitat for rare wild flowers, such as orchids, along with many species of bird including twite, dunlin, ringed plover and, when the cover is high enough, the often-heard but rarely seen corncrake.*

lochs near the west coast, the French-style Ormiclate Castle was built at the beginning of the 18th century as the residence of the Clan chief. By 1715, unfortunately, the castle had been destroyed in a fire, but its ruined walls can still be seen by visitors today. Later that century, South Uist became famous as a hiding place for Bonnie Prince Charlie who, aided by local girl Flora MacDonald, made his way back to France via Skye and Raasay after his failed bid to seize the British throne.

In 1838 South Uist, along with Barra and Benbecula, had been purchased by Colonel John Gordon of Cluny (see page 70). By 1851, around 2,000 crofters and their families had been forcibly evicted from South Uist to make way for large-scale sheep farming. These poverty-stricken folk were put on board ships in Lochboisdale and sent to Canada, where they struggled to start a new life. From a high of nearly 4,000 in 1841, South Uist's mainly Catholic and Gaelic-speaking population has dropped dramatically and now stands at around 1,800.

WALKING
On South Uist, as in the rest of the Hebrides, access to the majority of land is unrestricted. However, the public should always respect crofters' privacy, avoid walking on growing crops or entering certain bird breeding sites. Gates should always be closed and dogs should be kept under strict control.

Employment opportunities for islanders were greatly enhanced in 1958 when a missile-testing range was brought into operation by the MOD on the northwestern coast of the island. Along with facilities on neighbouring Benbecula and a tracking station on St Kilda the range, one of the best of its kind in the world, is now operated by QinetiQ to evaluate the new Eurofighter Typhoon's missile system.

NATURAL HISTORY

With its wide variety of habitats, ranging from 20 miles of continous sandy beaches, sand dunes and closely-cropped machair in the west to the hundreds of small inland lochs and rugged and remote mountains in the east, South Uist contains a wide variety of wildlife.

In the north of the island, Loch Bee provides a breeding ground for many species of wildfowl and is also home to around 500 mute swans. To the south is Loch Druidibeg, one of the most important National Nature Reserves in Scotland. Founded in 1958, its islets are one of the last strongholds of the scrubby woodland once found all over the Hebrides while the loch is an important breeding ground for resident greylag geese. Its 4,000 acres supports many birds

Left *The peace and tranquility of South Uist is portrayed in this view from South Glendale across the Sound of Eriskay to the causeway-connected island of Eriskay. South Glendale Primary School, now demolished, was the setting of Christina Hall's excellent book* Twice Around the Bay, *in which she tells the true story of her life as a teacher on South Uist in the 1940s and 1950s.*

COLONEL JOHN GORDON OF CLUNY

The eldest son of Charles Gordon of Braid and Cluny in Aberdeenshire, John Gordon was born in 1776 and, after receiving an excellent education, joined the Royal Aberdeenshire Light Infantry in 1800. He rose through the ranks before becoming an Honorary Colonel in 1836. Alongside his military career, he also found time to be MP for Weymouth and Melcombe Regis between 1826 and 1830. John Gordon eventually inherited many large family estates in Scotland as well as in the West Indies and, in 1838, purchased the Hebridean islands of Barra, South Uist and Benbecula. To make way for more profitable large scale sheep farms, Gordon soon set about ridding himself of the tenant farmers on those islands. Employing local militia and gangs of thugs, he forced thousands of destitute and starving crofters and their families on to ships with a one-way ticket to Canada.

The deserted townships on Barra, South Uist and Benbecula are not the only lasting memorials to this ruthless man. One of the original graffiti artists, during his Grand Tour of the Middle East in 1804-1805, Gordon carved his name on such wonders of the world as the Pyramids, the Dendara Temple, the Temple at Edfu, the Tomb of Paheri, the Temple of Esna, Gebel el-Silsila, the Temple at Karnak, the Pylon of Luxor Temple, the Temple of Medinet Habu, the Temple of Sethos I and on several tombs in the Valley of the Kings. We can be sure that history will never forget him! John Gordon died a wealthy man in 1858 and is buried in St Cuthbert's churchyard in Edinburgh.

Right *Much of the Outer Hebrides are formed of an ancient rock known as gneiss. Where this is found, the soil is acidic and unproductive but, in waterlogged areas, is overlaid by deposits of decayed vegetable matter known as peat. As well as an excellent indicator of climate change peat, in its dried form, is also an important source of fuel. Lines of drying peat, such as these near Crossdougal on South Uist, can still be seen throughout the Outer Hebrides.*

of prey including golden eagle, and a small area of rare woodland attracts migrant visitors during spring and autumn. The reserve also extends westwards to include the important machair habitat, with its colourful riot of wildflowers in the summer. These coastal areas are also important breeding grounds for many wading birds.

It is unusual to see a patch of green when looking at a map of South Uist! However, around the mouth of Loch Eynort on the island's east coast are several areas of native broadleaf woodland that have been reintroduced in recent years. This sheltered habitat now supports many woodland flowers as well as woodland and migrant birds not normally seen elsewhere on the island.

How to Get There
By sea Caledonian MacBrayne operate a regular vehicle and passenger ferry between Oban, Castlebay (Barra) and Lochboisdale (South Uist). CalMac also operate a regular vehicle and passenger ferry between northern Barra and Eriskay (for South Uist).

For more details on all of these services contact the CalMac reservations office (tel. 08705 650000) or visit their website: www.calmac.co.uk

By air British Airways operate regular daily flights from Glasgow to Barra and Benbecula. South Uist is accessible by road from Benbecula via a causeway. For more details contact BA (tel. 0870 850 9850) or visit their website: www.ba.com

Ordnance Survey Maps
Landranger 1:50,000 series Nos. 22 & 31

Tourist Information
Nearest office: Lochboisdale Tourist Information Centre, Pier Road, Lochboisdale, Isle of South Uist HS8 5TH (tel. 01878 700286) or visit website: www.visithebrides.com/islands/suist/

Where to Stay
There is a reasonable amount of accommodation on South Uist, ranging from two hotels to a good selection of bed and breakfast and self-catering establishments. For more details contact Lochboisdale Tourist Information Centre (see above) or visit the island's website: www.southuist.com.

Island Walks
With its low-lying machair, hundreds of small inland lochs and remote mountains, South Uist is an island of extremes. At over 20 miles long and 8 miles wide, there are numerous opportunities for walking, whether it be a stroll along the tracks of the western machair or a serious climb to the island's two eastern peaks of Hecla (1,988ft) and Beinn Mhor (2,034ft). Parts of the northern beaches are owned by the MOD as part of its missile testing range and walkers should be aware of red flags that signify when the range is in use. There is also good walking country around the Nature Reserve at Loch Druidibeg and along tracks along North Glen Dale around Lochs Chearsanais and Marulaigh in the extreme southeast.

Below *After many years of neglect, some of the traditional single-storey crofts in the Outer Hebrides are now being restored as comfortable holiday homes. Here, on South Uist on the southern shores of Loch Baghasdail, are views across to the ferry port of Lochboisdale and the rugged eastern half of the island.*

ERISKAY

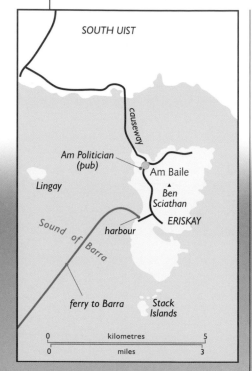

SOUTH UIST

causeway

Am Politician
(pub)

Am Baile

Lingay

Ben
Sciathan

ERISKAY

Sound of Barra

harbour

ferry to Barra

Stack
Islands

0   kilometres   5
0   miles   3

E riskay's poor soil probably saved it from a fate worse than death during the 19th century 'clearances', when displaced crofters from South Uist boosted the island's tiny population and turned it into a thriving centre for the herring fishing industry. After decades of terminal decline following World War II, this real-life *Whisky Galore!* island was thrown a lifeline in 2002 with the completion of a new causeway to South Uist and an improved ferry link with Barra.

HISTORY

Along with the neighbouring islands of Barra and South Uist, Eriskay was granted to the chief of the MacNeil clan by the Lord of the Isles in 1427. In 1838, however, General Roderick MacNeil became bankrupt and was forced to sell his lands to Colonel John Gordon of Cluny Castle. Gordon was one of the most ruthless landowners of the infamous 'clearances' but, mercifully, Eriskay escaped his excesses - but only because the land was too poor to support farming.

For centuries, fishing had always been the main source of income for the small island community, which numbered 80 in 1841. An influx of displaced crofters from South Uist saw the population rocket until it reached 466 in 1881, but by then Eriskay had also become a major player in

the Scottish herring fishing industry, a position it held for the next 50 years.

In the intervening years, Eriskay's scholarly Roman Catholic priest, Father Allan McDonald (who also built St Michael's Church) became internationally renowned for his work in preserving traditional Gaelic songs, poetry and folklore for future generations of Scots.

By World War II, the herring fishing industry was finished and the island's population went into decline, until it reached 133 in 2001. However, the opening of a new causeway to South Uist and improved ferry links to Barra in 2002 may have come just in time to save the dwindling fortunes of this lovely island.

NATURAL HISTORY

A riot of colourful wildflowers in the summer, Eriskay is probably best known for its unique breed of native pony, which has been used for centuries as a tough little work horse on crofts throughout the Hebrides. Usually coloured grey, but sometimes black or bay, they only stand 12.0–13.2 hands high and have a dense waterproof coat that allows them to live out in very harsh

Left *Mounted on a bracket in the grounds of St Michael's Church is the ship's bell from the World War I German battlecruiser SMS Derfflinger. Displacing 31,200 tons and launched in 1913, she saw action in the North Sea at the Battles of Dogger Bank and Jutland. Her crew scuttled her on 21 June 1919, after she had been interned at Scapa Flow.*

THAT ELUSIVE CORNCRAKE!
Once common throughout Britain, the corncrake is now only found on a few Hebridean islands such as Eriskay. They arrive from overwintering in Africa in April and May and prefer to spend their time hidden in tall vegetation such as irises, cow parsley or nettles. Sadly, modern farming practices such as early mowing for silage have destroyed much of their habitat, and they are now an endangered species in Britain. Although their distinctive rasping call can often be heard during the summer, corncrakes are rarely seen.

conditions. For more details about this unique animal, visit the Eriskay Pony society website: www.eriskaypony.com

HOW TO GET THERE
By road A new causeway opened in 2001 now links Eriskay with South Uist. (How to get to South Uist - see p.70)
By sea Caledonian MacBrayne operate a frequent vehicle and passenger ferry between Eriskay and northern Barra. For more details contact their reservations office (tel. 08705 650000) or visit their website: www.calmac.co.uk

Below *The white sandy beach at Coilleag a' Phrionnsa on Eriskay's west coast was where the exiled Bonnie Prince Charlie first set foot on Scottish soil, when he was landed here by the French frigate* Doutelle *on 23 July 1745.*

ORDNANCE SURVEY MAPS
Landranger 1:50,000 series No. 31

TOURIST INFORMATION
Nearest office: Lochboisdale Tourist Information Centre, Pier Road, Lochboisdale, Isle of South Uist HS8 5TH (tel. 01878 700286) or visit website: www.visithebrides.com/islands/suist/

WHERE TO STAY
Apart from a couple of self-catering establishments, there is currently no other accommodation on Eriskay. However, a wider choice is available on neighbouring South Uist (via the new causeway). For more details contact Lochboisdale Tourist Information Centre (see above).

Above *Built at a cost of £9.4 million (roughly £70,000 per head of the island's population) and opened on 11 September 2002, this brand new causeway now links Eriskay with South Uist., seen in the distance. New ferry terminals were also built on both Eriskay and northern Barra to complete the last link in the important communication chain between the islands of the Outer Hebrides.*

ISLAND WALKS
Less than 3 miles long and 2 miles wide, little Eriskay is ideal to explore on foot. Good views can be had of Barra and South Uist from the island's highest point, Beinn Sciathan (607ft). End your walk with a well-earned pint at the delightful *Am Politician* pub which has far-reaching sea views from its garden.

WHISKY GALORE!
True story: On 5 February 1941, the merchant ship *SS Politician* ran aground on the mainly Catholic island of Eriskay while en route from Liverpool to Jamaica and New Orleans. In her cargo she was carrying 28,000 cases of Scotch malt whisky, newly-minted Jamaican bank notes and £145,000 of British currency. The crew were saved and some of the whisky had already been 'liberated' by locals before a local customs officer and police arrived on the scene. Several islanders were arrested and subsequently jailed for theft, while the wreck of the ship and its remaining cargo was dynamited.
The film: Released in 1949 and based on the book of the same name by Compton MacKenzie, the Ealing comedy *Whisky Galore!* was produced by Michael Balcon and starred Basil Radford, Joan Greenwood, Wylie Watson and up-and-coming young actors James Robertson Justice and Gordon Jackson. In the film the ship becomes the *SS Cabinet Minister*, Eriskay becomes the Calvinist (dry) island of Todday and the amount of whisky becomes 50,000 cases. Using locals as extras who were paid £1 per day, *Whisky Galore!* was filmed on location on the island of Barra over a period of three months in the summer of 1948. Running well over its budget due to the dreadful weather, the film was titled *Tight Little Island* in the USA.

BARRA AND VATERSAY

Ancestral home of the MacNeil Clan, Barra and its predominantly Gaelic-speaking people suffered grave injustices at the hands of its ruthless landlord, Colonel Gordon of Cluny, during the 'clearances' of the mid-19th century. Since then, the MacNeils have reclaimed their estate and given most of it back to the people of Scotland. Both Barra and Vatersay, the latter owned by the government since 1908, are renowned for their abundance of rare wildflowers during spring and summer.

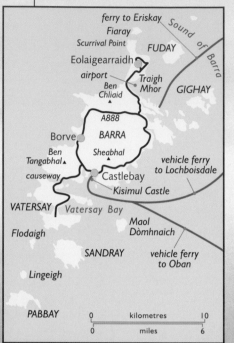

HISTORY

Archæological remains, such as standing stones and passage graves found on Barra and Vatersay, show that the islands have been inhabited for 6,000 years – since the New Stone Age. Both the Bronze Age, in the shape of circular stone cairns, and the Iron Age, in the shape of wheelhouses and brochs, are also well represented on the islands. In 2007, an archæological dig at Allasdale by the TV programme *Time*

Below *Oystercatchers take a break on an unspoilt beach near Borve Point on Barra's west coast. Common throughout the Scottish islands, the oystercatcher has a strong flattened bill which it uses to prise open cockles and mussels.*

Team uncovered 11 burial chests, including the well-preserved skeleton of a woman, which probably date back to 1,500BC. Following settlement by the Picts and the early Celtic Christians Barra, along with the rest of the region, came increasingly under Norse influence and did not become part of the kingdom of Scotland until the signing of the Treaty of Perth in 1266.

Along with the neighbouring islands of Eriskay and South Uist, Barra was granted to the chief of the MacNeil Clan by the Lord of the Isles in 1427. By the 17th century the MacNeils, who had expanded their empire to include the Bishop's Isles (see page 77), were a

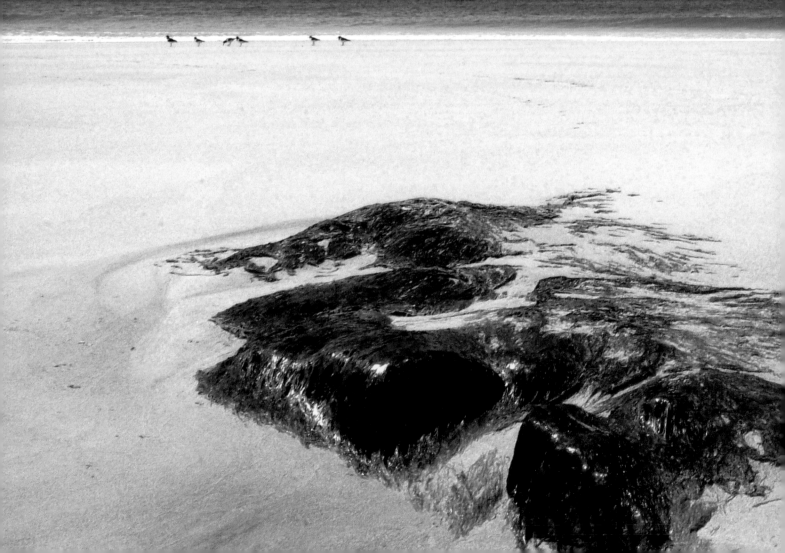

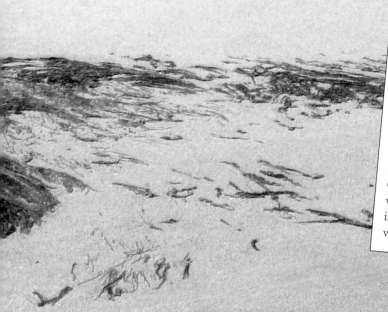

powerful force in the region and were not averse to a spot of piracy or cattle raiding. In 1838, however, despite investing heavily in larger farms and the fishing and kelp-burning industries, General Roderick MacNeil became bankrupt and was forced to sell his lands to Colonel John Gordon of Cluny Castle. Gordon was one of the most ruthless landowners of the infamous 'clearances' and, at the height of the potato famine between 1848 and 1851, he and his gangs forcibly removed about 1,200 people from Barra, many of whom were given a one-way ticket to Canada.

Above *Castlebay, the capital of Barra, is dominated by the impregnable 15th century Kisimul Castle and the 19th century Church of Our Lady Star of the Sea.*

The latter part of the 19th century saw an upturn in Barra's fortunes, with a boom in the herring processing industry and the passing of the Crofters Holdings (Scotland) Act of 1886, which gave the island's crofters fair rents and security of tenure. By 1908 the government had bought the island of Vatersay, along with land on Barra, to provide extra crofts for the hundreds of peasant farmers, or cottars, that lived on the islands.

The penultimate chapter in Barra's

fortunes occurred in 1932, when Colonel Gordon's descendant, Lady Gordon Cathcart, died. Her estates on Barra, including the long-abandoned Kisimul Castle, were bought by Robert MacNeil, an American who had traced his ancestry and proved that he was the 45th Chief of the Clan MacNeil. The last chapter came in 2004, when his son and current Chief of the Clan, Ian Roderick MacNeil, gave the crofting part of the MacNeil estate on Barra to the people of Scotland.

KISIMUL CASTLE

Strategically located on an island in Castlebay harbour, Kisimul Castle was the seat of the MacNeils of Barra from the 15th century until it was abandoned in the mid-18th century. Following bankruptcy in 1838, the MacNeils lost their lands on Barra but, in 1937, an American descendant, Robert Lister MacNeil, bought back much of the estate and spent the next 30 years restoring the castle to its former glory. In 2000, the present Clan Chief, Ian Roderick MacNeil, leased the castle to Historic Scotland. Open to the public, Kisimul Castle is also the venue for the worldwide gathering of the Clan MacNeil. For more information about the clan visit their website: www.the-macneils.org.uk

Above *On the regular service from Glasgow and Benbecula, a British Airways De Havilland Twin Otter comes in to land on the vast expanse of Traigh Mhor beach. Barra is the only airport in the world where scheduled flights land on a beach.*

Right *Testimony to the declining population of the last 150 years, the islands of the Outer Hebrides are littered with deserted and ruined houses. On Vatersay, this majestic example stands like a monolith overlooking Castle Bay and Barra. Vatersay was physically joined to Barra in 1990 with the opening of a new causeway.*

NATURAL HISTORY

With their wide range of habitats ranging from rocky inlets, shell-sand beaches and machair to the moorlands and uplands of their interior, unspoilt Barra and Vatersay are a wildlife paradise. In particular, Barra is renowned for its spectacular range of wildflowers during May and June, including 14 species or sub-species of the orchid family. Of these, the Irish lady's tresses is by far the rarest. The rare pink-flowered sea bindweed, also found on Eriskay, grows on Vatersay. Although the Outer Hebrides are mainly treeless, the stands of trees at Brevig and Northbay on Barra are regulary visited by rare migrant birds swept thousands of miles off course by strong westerly winds in the spring and autumn.

HOW TO GET THERE
By sea
Caledonian MacBrayne

operate a regular vehicle and passenger ferry between Oban and Castlebay on Barra. On Thursdays during the summer, this service also calls at Coll and Tiree. CalMac also operate a regular vehicle and passenger ferry between northern Barra and Eriskay (for South Uist). Certain ferries on the Oban to Lochboisdale (South Uist) service also call at Castlebay.

For more details on all of these services contact the CalMac reservations office (tel. 08705 650000) or visit their website: www.calmac.co.uk

By air British Airways operate regular daily flights from Glasgow to Barra and on to Benbecula. For more details contact BA (tel. 0870 850 9850) or visit their website: www.ba.com

ORDNANCE SURVEY MAPS
Landranger 1:50,000 series No. 31

TOURIST INFORMATION
Nearest office: Castlebay Tourist Information Centre, Main Street, Castlebay, Barra HS9 5XD (tel. 01871 810336) or visit their website: www.visithebrides.com/islands/barra/ or the island's website: www.isleofbarra.com

Above *Built as a three-storey dwelling in the 15th century for a member of the MacNeil Clan, this stone tower, known as Dun Mhic Leoid, stands on a small island in landlocked Loch Tangasdail.*

WHERE TO STAY
There is a good range of accommodation available on Barra, ranging from hotels and bed and breakfast establishments to self-catering units and a hostel. For more details contact Castlebay Tourist Information Centre or visit the island's website (see above).

ISLAND WALKS
With its rugged interior, Barra is an ideal destination for the hillwalker. Utilising the excellent bus service (Monday – Saturday) around the island's

coastal road, it is possible to plan several interesting walking itineraries without retracing one's steps. At low tide, the miles of shell-sands of Traigh Mhor in the north offer a chance to see one of Barra's oldest industries in action – collecting live cockles. The near-deserted island of Vatersay, with its beautiful white sandy beaches, is also served by a regular bus service (Monday – Saturday) from Castlebay.

THE BISHOP'S ISLES
The chain of uninhabited islands to the south of Vatersay – Sandray, Pabbay, Mingulay and Berneray – are collectively known as the Bishop's Isles. Apart from Sandray, they are all owned by The National Trust for Scotland. The most southerly of the Outer Hebrides, they are not only renowned for their dramatic cliffs, sea stacks and arches but also for their large seabird population. A deserted village on Mingulay is all that remains of a community of 150 people who, by 1912, had left the island, many of them settling on Vatersay. The southernmost island, Berneray, is the dramatic location for Barra Head lighthouse. Perched on a clifftop nearly 700ft above the sea, the lighthouse was designed by Robert Stevenson and completed in 1833. Once the storm-lashed home for three lighthouse keepers and their wives, Barra Head was automated in 1980. There are boat trips to the Bishop's Isles from the harbour at Castlebay.

ST KILDA

Formed from an extinct volcano, St Kilda is an archipelago of four islands and several sea stacks located about 40 miles out into the Atlantic west of Benbecula. In 2005, in recognition of both its natural beauty and habitats and its historical heritage, St Kilda became one of only 24 places in the world to be awarded Dual World Heritage Status by UNESCO. Owned by the National Trust for Scotland, St Kilda is definitely one of the top destinations that you must visit before you die!

HISTORY

Probably the most remote community in Britain, the island of Hirta – the largest of the St Kilda archipelago – was first visited by Bronze Age man around 4,500 years ago. Archæological remains found on the island also show that it was probably continually inhabited for at least 2,000 years until the last islanders left in 1930. Norse names for the island's hills, such as Ruaival, and Viking jewellery and weapons found on Hirta, also point to a continuing occupation.

Following the end of Norse rule in the region, St Kilda was eventually granted by the Lords of the Isles to the MacLeods of Dunvegan Castle in Skye. For centuries the hardy islanders, numbering around 180 until the 18th century, survived on fishing, growing barley and oats, raising sheep and, primarily, eating the local and plentiful seabird population. The islanders paid rent in kind (seabird oil, wool, etc) once a year to the factor of their absentee landlord in Skye.

Disease, in the form of smallpox and cholera, first decimated the population in

Above *Protected under the Ancient Monuments Act, the settlement on Hirta has been uninhabited since 1930 when the remaining 36 inhabitants were evacuated to the mainland.*
Below *The pattern of Hirta's settlement is best viewed from the summit of Ruaival (728ft). The modern military installation is conveniently hidden from view behind this stone cleit - one of around 1,400 that are dotted around the island and used for centuries as larders or stores by the hardy islanders.*

Above *Covered with thousands of nesting gannets, the cliffs of Boreray rise vertically out of the sea to a height of over 1,000ft. Landing on this island is fraught with danger and prior permission is required from the warden on Hirta. For centuries, St Kildans risked their lives on these cliffs, sometimes at night, to collect gannets and their eggs. The remains of three bothies on the island were used as shelter by these fearless men and women and by shepherds attending sheep.*

the mid–18th century, and life became increasingly harsh for the islanders. Many emigrated to Australia in the 1850s, leaving a much-depleted population of around 70 by the beginning of the 20th century. By then summer tourism, in the form of visiting steamers, was having a negative impact on the islanders' self-sufficiency and the decline continued until 1930 when the last 36 inhabitants, by now near-starving, asked to be evacuated to the mainland.

The islands were purchased by the Marquess of Bute in 1931 and finally bequeathed to the National Trust for Scotland in 1957. They, in turn, leased St Kilda to the Nature Conservancy (now Scottish Natural Heritage) as a National Nature Reserve. At about the same time, the army also moved in and built a radar-tracking station on Hirta as part of the new missile testing facilities based on South Uist. This facility has now been privatised, and is managed for the MOD by QinetiQ.

In recognition of their archæological and historical importance, the buildings on St Kilda have been protected under the Ancient Monuments Act since 1963 and, in 1987, became Scotland's first World Heritage Site.

Below *In the foreground, Stac Lee with its gannet colony rises from the sea to a height of 564ft. This stack, along with Stac an Armin Boreray, Hirta and Soay, the latter two seen in the distance, are the remains of a rim of a volcano that was active about 60 million years ago. Rising to a height of 1,411ft, the seacliffs on Hirta are the highest in the British Isles.*

Right *Gannet City! The world's largest colony of gannets, 60,000 pairs, breed on Boreray and the two adjacent stacks of Stac Lee and Stac an Armin. During the breeding season, every crevice of these sheer cliff faces is covered with nesting gannets, while the sky above witnesses the spectacular comings and goings of these magnificent birds. With a wingspan of over 6ft, gannets dive from a height of over 100ft, reaching speeds of 60mph as they plunge deep into the sea to catch fish for their young.*

NATURAL HISTORY

With a population of over one million birds, the archipelago of St Kilda is one of the most important seabird breeding sites in northwest Europe. On the sheer cliffs of Boreray and its stacks are around 60,000 pairs of gannet - a bird once vital to the islanders' diet and now the world's largest colony. St Kilda also boasts the largest colonies of puffin (140,000 pairs) and fulmar (62,000 pairs) in Britain. The islands are home to their own unique sub-species of wren, fieldmice and sheep - the prehistoric Soay sheep are thought to have been introduced by Bronze Age farmers around 4.500 years ago. On land, around 140 species of plants have been recorded while the clear, unpolluted waters around the islands are rich in marine life.

In recognition of its natural importance St Kilda, leased from the National Trust for Scotland by Scottish Natural Heritage, has been designated a National Nature Reserve, Biosphere Reserve, National Scenic Area, Site of Special Scientific Interest, World Heritage Site and a European Community Special Protection Area.

HOW TO GET THERE

By sea Not the easiest of journeys, but well worth the effort. There are several boat operators that visit St Kilda, mainly from the Western Isles. For more information visit the island's website: www.kilda.org.uk

Probably the shortest journey (about 41 miles each way) for a day trip is operated by Kilda Cruises from Leverburgh in South Harris. For more details contact Angus Campbell (tel. 01859 502060) or visit their website: kildacruises.co.uk *Author's note:* It is advised that travellers take precautions against seasickness!

ORDNANCE SURVEY MAP

Landranger 1:50,000 series No. 18

TOURIST INFORMATION

Nearest office: Tarbert Tourist Information Centre, Pier Road, Tarbert, Isle of Harris HS3 3DJ (tel: 01859 502011) or visit the website: www.visithebrides.com

WHERE TO STAY

Apart from working as a volunteer for the Scottish National Trust (www.kilda.org.uk) there is no accommodation on St Kilda. The nearest accommodation is on South Harris. For more information contact Tarbert Tourist Information Centre (see above).

ISLAND WALKS

Visitors to St Kilda are advised to be prepared for extremes of weather and wear suitable clothing and footwear. On arrival they are met at the landing stage by the resident warden who will advise on appropriate walking routes. Walking around the old village is fairly level but the rest of the island can involve strenuous climbs. Please follow the warden's advice on avoiding sensitive seabird breeding areas. *Author's note:* It is a most unpleasant experience to be attacked by a flock of angry great skuas!

ROCKALL

The plug of an extinct volcano, located 187 miles west of St Kilda, Rockall is the westernmost part of the United Kingdom. Rising to a height of 75ft, this uninhabited islet was claimed as part of the UK as recently as 1955 when a Royal Navy party were landed by helicopter and raised the Union Jack. Located in an area of potential major oil exploitation, Rockall has in the past been the subject of many territorial claims from Ireland, Iceland and the Faroe Islands (Denmark).

Below *On Islay's inaccessible east coast, the mountain peaks of Beinn na Caillich and Sgort nam Faoileann tower over the lighthouse on McArthur's Head. Guarding the southwest entrance to the narrow Sound of Islay, the lighthouse was built in 1861 and became automated in 1969. It can only be reached on foot along a walking trail from Ardtalla.*

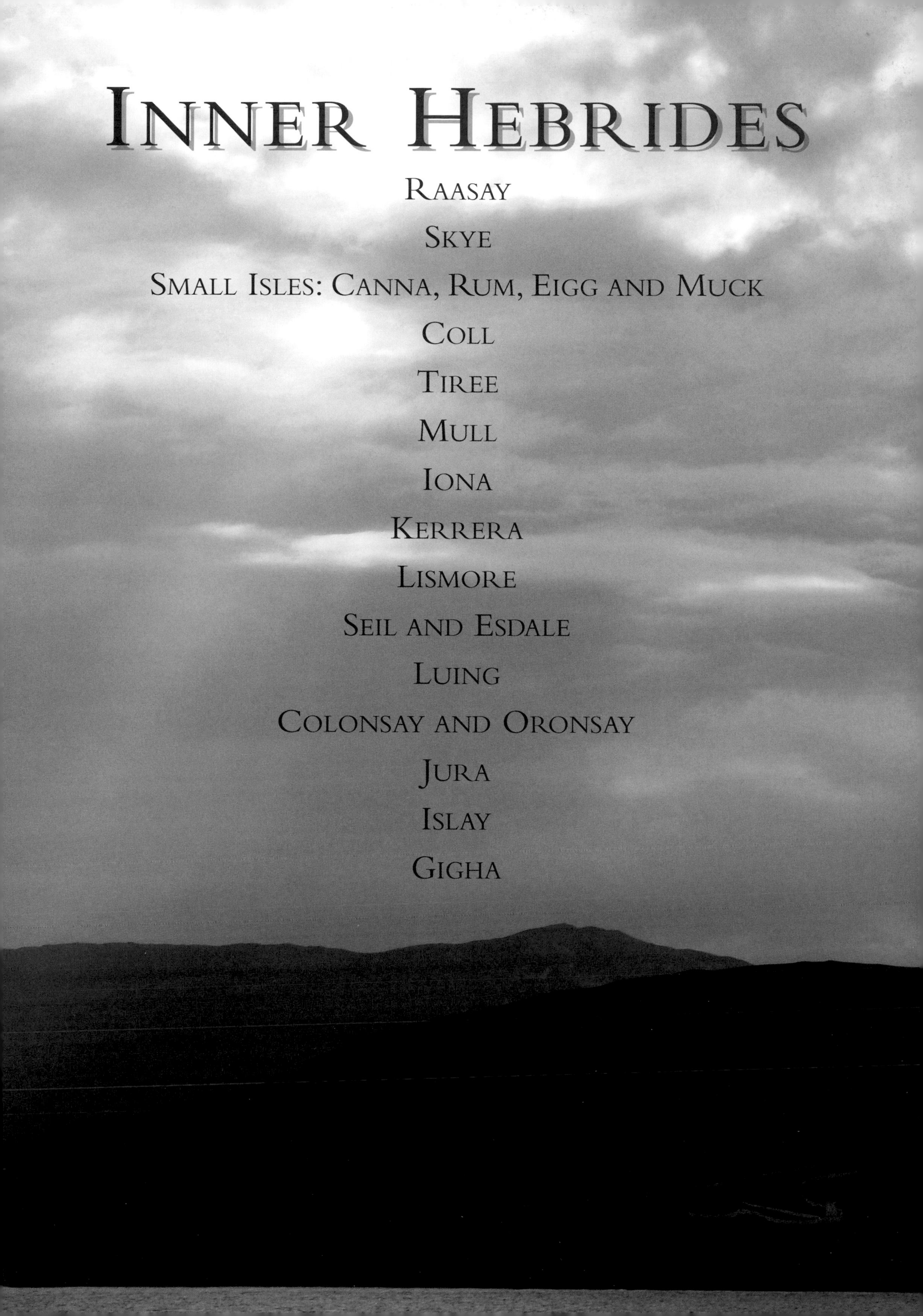

INNER HEBRIDES

Raasay

Skye

Small Isles: Canna, Rum, Eigg and Muck

Coll

Tiree

Mull

Iona

Kerrera

Lismore

Seil and Esdale

Luing

Colonsay and Oronsay

Jura

Islay

Gigha

RAASAY

From the early hunter-gatherers of the Stone Age and centuries of Viking domination to 350 years of rule by the same family and mass emigration during the 19th century, the little island of Raasay has its own unique and rich history. Still carrying the scars of a short-lived early-20th century iron ore mining enterprise, the island and its much reduced population now look to forestry, farming and tourism as their main economic activities.

Below *The old pier and boathouse at Clachan are located close to Raasay House (now an Outdoor Centre) overlooking Churchton Bay. A defensive battery was built next to the boathouse in 1809 during the threat of invasion from France.*

HISTORY

From the rock shelters and stone tools of early Stone Age man and burial mounds of the Bronze Age to the brochs and underground passages of the Iron Age, there is a wealth of archæological evidence to show that Raasay has been continually inhabited for nearly 10,000 years. Following visits by the early Christians in the 6th century, the region had come under the control of the Vikings by the 9th century – the name 'Raasay' is thought to be Norse for 'Roe Deer Island' – and, for centuries, the fortunes of Raasay were closely linked with those of its larger neighbour, Skye (see pages 86–93). Religious influence reasserted itself in the late 12th century when a large chapel, dedicated to St Moluag, was built on Raasay. .

From the early 16th century until 1846, when it was sold to George Rainy, a wealthy London merchant, Raasay was ruled by the MacLeods, first from Brochel Castle and later from Raasay House. During the 1745 Rebellion, the MacLeods supported Bonnie Prince Charlie's claim to the throne of Scotland. After the rebellious Scots were defeated at Culloden in 1746, however, the Prince sought refuge on Raasay from government troops who were searching for him. Although he escaped capture and fled to France, the islanders received the full wrath of the government troops with almost all of their homes, livestock and boats being destroyed.

Raasay had certainly recovered from its part in the Rebellion when, in 1773, it was visited by Dr Samuel Johnson and James Boswell on their epic trip around the Hebrides. For three days they were treated like visiting royalty, and the pair were very impressed by what they were shown. Unfortunately, life for the islanders was soon to become much harder.

Although new industries were introduced such as fishing, kelp burning and forestry, the extravagant lifestyle of the MacLeods became a drain on the island's economy. With crops failing and the price for kelp and cattle plummeting, many islanders soon faced destitution. The first wave of emigrations, to Canada, began in the early 19th century and had come to a head by the 1850s and 1860s, when many more left for Australia. Since

ISLE OF SCALPAY

To the south of Raasay lies the privately-owned Isle of Scalpay. A large island, its rugged interior (Mullach na Carn is the highest point at 1,299ft) contains several small lochs and a resident red deer population. With an extensive range of habitats, including 200 acres of woodland and 14 miles of coastline Scalpay is also home to a wide variety of animal and plant species. There are three self-catering cottages available to rent on the island. For more details visit the island's website: www.isleofscalpay.com

Right *Located on the northeast coast of Raasay, Brochel Castle was built in the 15th century and was used as a base and then a home for the MacLeods until 1671.*

then, the population of Raasay has fallen from a high of 900 to its current level of just under 200.

During the 19th century, Raasay changed hands several times until the island was bought in 1912 by William Baird & Co to mine iron ore deposits. A new pier, railway and kilns were built and, by 1914, the industrial revolution had started in earnest on Raasay. Shortages of local labour during World War I led to German prisoners of war being employed in the mines but, due to falling world prices for ore, by 1920 they had closed. Today, visitors to Raasay are greeted by the ghosts of this short-lived venture in the shape of the old pier, iron ore hopper and kiln bases.

Raasay was sold to the Board of Agriculture for Scotland (later to become the Department for Agriculture & Fisheries) in 1923 and the land let back to crofters. Parts of the island, including Raasay House, were sold to absentee landlord, Dr John Green, in 1961. Following a long acrimonious period when he refused to sell land for a new ferry terminal, he was finally bought out by the Highlands & Islands Development Board in 1979.

NATURAL HISTORY

Ranging from managed mixed woodland, lush pastures and heather moors to small inland lochs and a wild eastern coastline, the island of Raasay supports a wide variety of natural life including red deer, otter, pine marten, the unique Raasay vole and over 60 species of bird. Porpoise and minke whale are often seen in the waters of the Inner Sound to the east of Raasay. From spring to summer, visitors are greeted by successive displays of bluebell, rhododendron and a wide variety of other flora including many rare species of wild orchid.

HOW TO GET THERE

By sea Caledonian MacBrayne operate a vehicle and passenger ferry between Sconser on the Isle of Skye to Raasay. For more details contact their reservations office (tel. 08705 650000) or visit their website: www.calmac.co.uk

ORDNANCE SURVEY MAPS

Landranger 1:50,000 series No. 24

TOURIST INFORMATION

Nearest office: Portree Tourist Information Centre, Bayfield House, Bayfield Road, Portree, Isle of Skye IV51 9EL (tel. 01478 612137) or visit the VisitScotland website: www.visitscotland.com

WHERE TO STAY

In addition to the 12-bedroom Isle of Raasay Hotel, there is the Raasay House Outdoor Centre, a Youth Hostel and a small number of self-catering and bed and breakfast establishments on the island. For more details of these visit the island's website: www.raasay.com

Alternatively, contact Portree Tourist Information Centre or visit the VisitScotland website (see above).

ISLAND WALKS

Around 14 miles long and 3 miles wide and with stunning coastal scenery and minimal traffic, Raasay is well worth exploring on foot. An excellent illustrated free guide to 17 walks on the island has been produced jointly by the Forestry Commission and Raasay Social Services Association.

ISLE OF SKYE

A wild and dramatically beautiful island, Skye has magnificent ruined castles, splendid sea lochs, high mountains, remote moors and steeply falling waterfalls. It offers plenty for rock climbers, birdwatchers and walkers alike. The island is known in Gaelic as Eilean a' Cheò – the 'Misty Isle' – and is the largest island of the Inner Hebrides as well as the second largest island in Scotland after Lewis with Harris. There are 350 miles of coastline to explore and, at its heart, the granite ridge of the Cuillin Hills rises magnificently to over 3,000ft.

Below *One of the most beautiful routes in Skye is the road from Broadford to Elgol. Here, the tall marsh grasses bend to the wind on Loch Cill Chriosd - named after the nearby medieval church and its graveyard.*

HISTORY

Skye has had a turbulent past, with clan wars and rebellion both leaving their mark. On this fascinating and varied island, there is evidence of settlements that date back to 6500 BC. The early inhabitants left a legacy of standing stones, cairns, brochs, gravestones and monastic sites. St Columba visited Skye in 585AD.

The Vikings settled and dominated Skye for nearly 400 years from 794AD. A Viking fortress stood on the western edge overlooking Kyle Hakon, or Hakon's Strait, which separates the island from the mainland. The ruin of the 15th-century Castle Moil, stronghold of the MacKinnons, now stands there. It is said to have been built by a Norwegian princess, known locally as Saucy Mary, who ran a chain across the straits and would only let boats through if they paid a toll.

In 875, Norwegian settlers began to arrive, and there were frequent clashes with Norway over the following centuries. Under the rule of the Celtic Lords of the Isles, Skye maintained its independence from the Scottish kings

Right *Cill Chriosd (Christ's Church), located two miles from Broadford on the long and winding road to Elgol, was built in the late 15th century. In 1627 the church received its first Protestant minister, Neil MacKinnon. The church was superceded in 1840 by a new parish church in Broadford.*

until 1263, when Alexander III won the Battle of the Largs. This ended Norse power on the island, but the Lords of the Isles presided over many more battles before the end of the 15th century, when their power was broken by James IV.

There were many attempts to crush the clan system and reduce the power of the chiefs. Clan disputes were common on Skye until well into the 17th century. In 1578, Trumpan Church near Uist became the setting for one of the most terrible events in clan warfare in Scotland. On the first Sunday in May, a raiding party of MacDonalds from Uist trapped the local MacLeods in the church and set fire to it, burning all those inside to death. One small girl escaped and ran for help to Dunvegan, where other MacLeods gathered, marching to Trumpan and killing the MacDonalds who were trapped there, stranded by the receding tide. All the dead were buried beneath a toppled wall.

It was following the Jacobite rebellion and their defeat at the Battle of Culloden in 1746 that the island

welcomed its most famous visitor, Prince Charlie Stuart – Bonnie Prince Charlie – had fled the battlefield and gone into hiding with a price of £30,000 on his head. On Benbecula, he was introduced to 23-year-old Flora MacDonald and she smuggled him to Skye, dressed as her serving maid, Betty Burke, under the noses of Hanoverian soldiers and bounty hunters. They landed at Portree, now

capital of Skye and the prince hid in a cave while Flora went to get help. While he was on island, the prince shared a secret recipe for a liqueur – now known as Drambuie but no longer brewed on the island – with the MacKinnon family who lived in an inn at Broadford Bridge.

Bonnie Prince Charlie escaped on June 26 to the island of Raasay, from whence he took ship

ISLE OF RONA

This island lies off Trotternish in the Inner Sound between Skye and the mainland. It is uninhabited now, apart from one house and a few holiday cottages. The island is a mile and a half long, and has one 300ft hill. Once a year, shepherds from Lewis round up the half-wild sheep. Naturalists have a field day here – kittiwake, guillemot, great skua and black-backed gull all breed here, and grey seal and killer whale can be seen in the waters around the island. A treat for ornithologists is a rare bird, the Leach's petrel, that lives out at sea, but comes ashore at Rona to breed, nesting in burrows.

for France. Before he left Skye, he said goodbye to Flora at McNab's Inn, where the Royal Inn now stands in Portree, giving her a locket containing his portrait – they never met again.

Flora's adventure was not over, however. She was arrested and imprisoned first in Dunstaffnage Castle, then on a prison ship, and finally, briefly, in the Tower of London. She was given amnesty in 1747 and married Allan MacDonald before emigrating with him to America until 1779. The couple returned to Skye and she died there in 1790, in the bed in which the prince had slept. Three thousand people attended their heroine's funeral, and she was buried in the churchyard at Kilmuir, her grave marked by a tall Celtic cross. In 1773, Dr Johnson wrote of her that she "will be mentioned in history, and if courage and fidelity be virtues, mentioned with honour."

The people who lived on Skye proved irrepressible. In 1886, after an uprising

Below *The magnificent view from Elgol harbour at the tip of the Strathaird peninsula across Loch Scavaig to the Cuillin Hills. Reaching heights of over 3,000ft these jagged peaks are considered by many to be the most dramatic range of mountains in Britain. The remnants of an volcanic eruption over 50 million years ago the Cuillins attract walkers and climbers throughout the year.*

and a subsequent Act of Parliament, Glendale became the first area where crofters were given security of tenure and the right to own their title deeds. During the 19th century, the population of Skye reached an all-time high of 23,000. In Victorian times, mountaineering, particularly in the Cuillin Hills, became very popular.

Today, just over 9,200 people live on Skye, but the population is growing, unlike many other Scottish islands. The main industries are tourism, agriculture, distilling whisky, brewing and crafts. New industries include potash from seaweed, used in the manufacture of glass and soap.

There are many traditional aspects to life on Skye: shinty, a team sport played with sticks and a ball, is still played regularly, and a strong folk music culture means that the annual Isle of Skye Music Festival is increasingly popular. The island is also home to the world-renowned MacCrimmon pipers, hereditary pipers to the MacLeod clan chiefs for 13 generations. The island is soaked in the Gaelic culture and heritage. Half the population still speak Gaelic and students of the language travel from all over to study at the Scottish Gaelic College in Sleat. The poet Sorley MacLean, born on the Isle of Raasay off Skye's east coast, lived most of his life on Skye. The oldest breed of terrier in Scotland, the Skye

Terrier, was possibly bred from dogs that survived a wrecked man-of-war from the Spanish Armada in 1588 and were mated with local terriers to produce a dog with a long, silky coat.

Natural History
Skye is about 50 miles long and varies in width between 7 and 25 miles. The coastline is so indented that there is no point that is more than 5 miles from the sea.

There are five peninsulas – Duirinish, Minginish, Trotternish, Waternish and Strath. Violent volcanic activity over one million years ago have created a dramatic landscape. The ice ages have formed U-shaped glens and wonderful mountainous peaks.

The Cuillin Hills, also called the Black Cuillin, dominate the skyline particularly from the far side of Loch Scavaig. They rise to more than 3,000ft and provide some of the most dramatic mountain terrain in Scotland. Birdwatchers can see goldcrest and golden eagle, sea eagle, gannet, cormorant and the black guillemot. There are many wild flowers because there is such a wide range of geological features. Mammals include pygmy shrew, red deer and, particularly, otter. Skye was home to the author Gavin Maxwell for the last year of his life. He lived on a lighthouse island at Kyleakin, and the island was the inspiration for his story about an otter in 'Ring of Bright Water'. The Otter Haven at Kylerhea in the southeast is testimony to the abundance of this entertaining mammal on Skye. There is a hide from which the visitor can also watch eagle, falcon, buzzard, seal and seabirds.

Above *A CalMac ferry leaves the harbour at Uig on the west coast of the Trotternish peninsula. This regular ferry service to the Outer Hebrides links Skye with Tarbert on Harris and Lochmaddy on South Uist.*

In the northwest of Skye at Waternish, are sea cliffs and headlands, black coral shores, deep lochs and high mountains. At Idrigill Point on Loch Bracadale there are pinnacles called MacLeod's Maidens rising from the sea. The water is very clear, so whale-spotting is an attraction. Dunvegan, the seat of the Clan MacLeod is at the heart of this area dominated by

Below *Located on the Trotternish peninsula, The Quiraing is one of the most unusual landscapes on Skye. With names like The Table, The Prison and The Needle, these rocks were once used by farmers as shelter for their sheep and cattle.*

the Cuillin Hills. The basalt has produced flat-topped shapes called MacLeod's Tables. It is said that these were named for a chieftain who entertained lairds from the mainland who wagered that he could not provide a larger table than that in the king's court in Edinburgh. MacLeod had a feast laid up there in the snow. There is also a flat-topped island offshore, visible from the ruin of Dunthulm Castle on the northern headland, that is called Lord MacDonald's Table.

Trotternish, the peninsula in the northeast, is the most northerly part of Skye and is dotted with alpine flowers. The Quirang, a large rock formation in the north, has the largest landslips in Britain. These happened because the weight of the basalt formed from lava flows cooling is too much for the rock structures underneath, causing some spectacular collapses over the centuries. Kilt Rock, on the eastern coast, is called that because the black basalt has folds that look like the pleats of a kilt.

Further south on Trottenish, there is a wonderful ridge for walkers, with Storr at its highest point. The Old Man of Storr is the tallest of the pinnacles here, a basalt column 160ft tall and 40ft around. Other basalt pinnacles nearby are called the Old Man's Wife, the Castle, and the Dog. This landscape is full of many different and strange rock formations and landslips. Kilt Rock overlooks the river that drains Loch Mealt and is the shortest in Scotland. It flows for only 150ft, than travels down the Mealt Falls, 300ft to the sea below. The falls are spectacular, not least because the water is often blown uphill by the powerful offshore winds.

The peninsula is famous for its plants of great botanical interest. The limestone outcrop of the ridge provides colourful plantlife throughout the year, from celandines and primroses in spring, through the meadowsweets and sneezeworts, yellow rattle and zigzag clover of summer to the knapweeds and scabious of autumn. The four varieties of Scottish thistle, the Spear, Creeping, Marsh and Melancholy, all thrive here. On the wetter heather and grass moorlands, all nine species of insect-eating plants native to Britain flourish. The Ridge itself is home to arctic alpine species, including the moonwort and the tiny Iceland purslane, which was discovered there in the 1950s.

The jagged Cuillin Hills dominate central Skye, and there are many places where wonderful views of its vertical ridges can be seen. Blà Bhein, or Blaven, often described as the most beautiful mountain in Scotland, is the most eastern peak of the Black Cuillin, separated from the main Cuillin range by Glen Sligachan. It is the highest of the mountains at 3,044ft, and can be a dangerous place with sudden changes in the weather. The Red Hills, also known as the Red Cuillin, provide a gentle contrast to the Black Cuillin because they are granite and more rounded.

The Sleat peninsula is unique on the Isle of Skye and is as varied as its geology. On the eastern side there is 3,000 million year old gneiss (the oldest rock in Britain), in the middle, brown sandstone and on the west side limestone and quartz. Known as the 'Garden of Skye', it is very different from the rest of the island, with its woodlands and fields. There are banks of rhododendrons and carpets of bluebells in spring, and abundant wild flowers in the pastures during the summer.

HOW TO GET THERE
By road Across the recently opened Skye Bridge on the A87 from Kyle of Lochalsh.
By sea Caledonian MacBrayne operate a regular vehicle and passenger ferry between Mallaig on the mainland and Armadale on the Sleat peninsula in southwest Skye. They also operate regular vehicle and passenger ferries between Tarbert (Harris) and Lochmaddy (South Uist) to Uig on the east coast of Skye. For more details on all of these services

DUNVEGAN CASTLE
Inhabited by the MacLeod clan for more than seven centuries, Dunvegan Castle is the oldest castle in Scotland to have been occupied continuously by the same family. On the edge of Loch Dunvegan in northwest Skye, it is the seat of the MacLeod of MacLeod, chief of the main part of the clan. It houses the Fairie Flag of Dunvegan, a magical flag said to have been given to the clan by the queen of the fairies because of a favour that the family did for her. It could be waved three times in time of need, and the fairies would help. It is said to have been waved twice already. The flag is made of silk from the Middle East and has been dated somewhere between the 4th and 7th centuries ad. It may originally have belonged to Harald Haardrada of Norway, defeated by Harold of England in 1066.

contact the CalMac reservations office (tel. 08705 650000) or visit their website: www.calmac.co.uk

ORDNANCE SURVEY MAPS
Landranger 1:50,000 series Nos. 23, 32 & 33

TOURIST INFORMATION
Portree Tourist Information Centre, Bayfield House, Bayfield Road, Portree, Isle of Skye, IV51 9EL (tel. 01478 612137) or visit the island's website: www.skye.co.uk

WHERE TO STAY
There is a wide range of all types of

accommodation on Skye. For more details contact Portree Tourist Information Centre or visit the island's website (see above)

ISLAND WALKS
The Isle of Skye is a paradise for walkers. The island can be divided into five main walking regions: Portree, Trotternish and the Braes; Dunvegan, Duirnish and Waternish; The Cuillin and Minginish; Broadford and Elgol; Kyleakin, Armadale and Sleat. For full details of walks in these regions, ranging from easy-going to serious climbing, visit the excellent website: www.skyewalk.co.uk.

Right *With its superb natural harbour, Portree is the administrative centre of Skye. Its name in Gaelic is Port-an-Righ which translates as 'King's Port' which dates back to 1540, when James V of Scotland visited the town. Once known as MacNab's Inn, the present Royal Hotel is where Bonnie Prince Charlie said farewell to Flora MacDonald in 1746.*

Below *The small inlet on Skye's east coast known as Loch Ailort is dwarfed by the peaks of Sgurr Mhairi (2,540ft) and Beinn Dearg Mhor (2,398ft). The loch is the location for a fish farm - one of many around the coastline of Skye - and an important contributor to the island's economy.*

TALISKER WHISKY
On the west coast of Skye, on the Minginish Peninsula, 21 underground springs rise into peaty burns that flood down Hawk Hill to Carbost, the home of Talisker Whisky. The distillery was built in 1830 by two doctor's sons, Hugh and Kenneth MacAskill, on the shores of Loch Harport. It produces a single malt whisky using five traditional wooden fermentation vats or washbacks, that are filled at the rate of 20,000 gallons of burn water an hour. It is double-distilled and this is the only place on Skye where malt whisky is made. Talisker was one of the whiskies feted by Robert Louis Stevenson in his poem, *The Scotsman's Return from Abroad*, written in 1880: "The king o'drinks, as I conceive it, Talisker, Islay or Glenlivit."

THE SMALL ISLES: CANNA

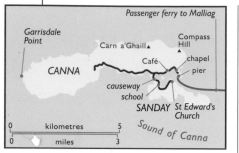

Following centuries of clan rule and mass evictions during the 19th century, the people of Canna finally found peace in 1938 when their island was bought by Gaelic scholar and historian, John Lorne Campbell. Gifted to the National Trust for Scotland in 1981, urgent steps have recently been taken to halt not only the decline in Canna's human population, but also its internationally important seabird colonies.

HISTORY

The remains of several Iron Age hut circles and coastal forts at the western end of Canna show that the island has been occupied for at least 3,000 years. By the 6th century, early Christian monks had arrived from Iona but were soon evicted in the late 8th century by Viking raiders, followed by centuries of rule from Norway. Canna, along with the rest of the Western Isles, only became part of the Kingdom of Scotland after the signing of the Treaty of Perth in 1266.

During the 13th century Canna, along with Eigg (see p.98), had come under the control of the MacDonalds of Clanranald. It is thought that Coroghon Castle, at the eastern end of the island and now ruined, was built by the MacDonalds in the late 17th century and used as a prison. Canna was sold to the MacNeills in 1827 and the new owner wasted no time in forcibly evicting hundreds of crofters and their families – the population had fallen from a high of 436 in 1821 to 127 by 1861.

Canna changed hands again in 1881 when it was sold to a wealthy Glaswegian businessman and then, in 1938, to the Gaelic historian and scholar, John Lorne Campbell. He gave the island and his Gaelic archives to the National Trust for Scotland in 1981 and continued to live in Canna House until his death in 1996. His wife, American musician Margaret Fay Shaw, continued to live on the island until her death in 2004. Sadly, Canna's population has continued to drop - by 2001 it had fallen to an all-time low of 12 - and the National Trust for Scotland recently advertised for new tenants to help revitalise the island. This, together with recent new pier facilities and improved links to the mainland, may protect Canna's long term future.

NATURAL HISTORY

Canna, once justifiably famous for its seabird colonies that nested along the precipitous western cliffs, is now one of 48 Special Protection Areas in Scotland. However, due entirely to the predatory brown rat which eats both their eggs and chicks, seabird numbers have fallen dramatically in recent years to dangerously low levels. Apart from the Manx shearwater, which has now more or less disappeared from the island, shag, herring gull, puffin and razorbill have all seen their numbers drop by at least 50%.

In a programme to eradicate the thousands of brown rats on Canna, the National Trust for Scotland placed over 4,000 poison bait stations around the island and, by 2006, the entire rat population had been destroyed. Prior to the eradication programme, the Royal Zoological Society of Scotland caught and removed to a temporary home over 150 of Canna's unique species of woodmouse. These have now been reintroduced to their island home, none-the-worse for their adventure.

HOW TO GET THERE

By sea Caledonian MacBrayne operate a vehicle and passenger ferry between Mallaig and the Small Isles. For more details contact their reservations office

Above *Close to the harbour, Canna's tiny protestant church, complete with intricately decorated iron gates, was built in 1914 in memory of Robert Thom, who bought the island in 1881.*

Below *A perfect glasshouse for growing indoor tomatoes, the island's only telephone box waits patiently for its next customer. On the headland in the distance are the ruins of 17th century Coroghon Castle and, to the right, Canna's tiny protestant church.*

Right *Nestling beneath the basalt sills and wooded slopes of eastern Canna, Canna House was once the home of the Gaelic historian and scholar Dr John Lorne Campbell, who gave the island to the National Trust for Scotland in 1981. To the right is the Harbour View Tearoom, the island's only refreshment stop.*

(tel. 08705 650000) or visit their website: www.calmac.co.uk

Day trips are also operated from Arisaig during the spring and summer by Arisaig Marine (tel. 01687 450224).

ORDNANCE SURVEY MAPS
Landranger 1:50,000 series No. 39

TOURIST INFORMATION
Nearest office: Mallaig Tourist Information Centre, East Bay, Mallaig PH41 4QS (tel. 01687 462170). Website: www.road-to-the-isles.org.uk

WHERE TO STAY
There is a limited amount of self-catering accommodation on Canna. For more details contact Mallaig Tourist Information Centre (see above) or the National Trust for Scotland who own the properties (tel. 0844 4932100), or visit their website: www.ntsholidays.com/coastlineandislands

SANDAY

Connected to Canna by a road bridge, low-lying Sanday is not only home to the only primary school on the two islands but is also the site of several Viking graves and the rather grand, but redundant, St Edward's Roman Catholic church.

ISLAND WALKS
Apart from the odd Landrover, quad bike or tractor, there are no vehicles or even tarmac roads on Canna. Walking is the only way to get around this little island and its neighbour, Sanday. As Canna is run as a farm, visitors are asked to keep to tracks and close gates. A favourite walk for visitors is to Compass Hill

(469ft) at the eastern end of the island from where there are far-reaching views of the surrounding islands on a clear day. Care must be taken along the coastline where clifftops end in a sheer drop to the sea, hundreds of feet below.

THE SMALL ISLES: RUM

Created over 50 million years ago in an age of massive volcanic activity, the wild and untamed Isle of Rum is by far the largest of the Small Isles. Inhabited for over 7,000 years, the island and its people have witnessed centuries of upheaval: from Viking raiders and clan feuding, to the mass evictions of the early 19th century, before becoming a playground for the Edwardian rich. Due to the importance of its diverse range of wildlife, Rum is now a National Nature Reserve.

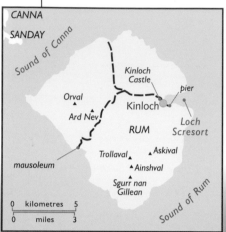

HISTORY

Archæologists have recently discovered remains on Rum that show that the island was occupied by early hunter-gatherers over 7,000 years ago. It is also thought that a type of hard rock found in the west of Rum, known as bloodstone, was an important source for tool making during the Stone Age. Bronze Age man left more visible remains in the shape of burial cairns at Kilmory and Harris and a standing stone near Laimrhig. The Iron Age is marked by two coastal forts on the northwest coast.

The first Celtic Christian monks probably arrived on Rum in the 7th century and several stone crosses from this period can be seen in the Old Burial Ground at Kilmory. From the 9th century Rum, along with the rest of the Western Isles, came under the influence of the Vikings, who kept their hold on the region until 1266 when the Treaty of Perth was signed. Rum then became part

Right *The twin peaks of Askival (2,664ft) and Hallival (2,372ft), formed during a period of major volcanic activity over 50 million years ago, dominate the wild southeastern corner of Rum known as the Cuillins. One of the most important wildlife sites in Scotland, Rum was designated a National Nature Reserve in 1957, a Biosphere Reserve in 1976, a Special Protection Area for Birds in 1982 and a Site of Special Scientific Interest in 1987. On the headland in the foreground is the ruined village of Port-na-Caraneon, which was abandoned after the 'clearances' of the 1820s.*

of the kingdom of Scotland and its population was often at the mercy of feuding clans, until it was acquired by the MacLeans of Coll in the 17th century. By this time, Rum's population had grown to around 200 people scattered around the island in several townships such as Port-na-Caraneon, Kilmory and Harris. With very little fertile land on the island, the main occupation of the islanders was fishing until sheep farming was introduced in the early 19th century.

Despite much poverty, Rum's population continued to grow until it had reached 443 by 1807. Sadly, life as they knew it, however hard, was about to end for most of the islanders. In the late 1820s, Rum was leased by the MacLeans as a large-scale sheep farm and the majority of the island's population were forcibly evicted. Most of these displaced people were put on board a ship bound for Nova Scotia in Canada.

The sheep farming venture was not a success and the MacLeans sold Rum in 1845 to the Marquis of Salisbury, who turned the island into a sporting estate for wealthy Victorians. It was sold again in 1870 and the shooting rights were subsequently leased by its new owner, F. Campbell, to the Lancashire MP and millionaire, John Bullough. Bullough loved the island so much that he, in turn, bought it from Campbell in 1888. Sadly, Rum's new owner did not enjoy his new possession for long as he died in 1891, leaving both the island and much of his wealth to his 21-year-old son, George.

George Bullough died in 1939 but his wife, Lady Monica, continued to live in the castle until 1957 when she sold Rum, Kinloch Castle and its contents to the Countryside Commission for Scotland for a nominal sum. Since that date, the island has been designated a National Nature Reserve, now managed by Scottish

Natural Heritage, and is an important site for ecological research. The Bullough family are buried in a lavish Greek temple-style mausoleum built by Sir George overlooking the sea at Harris in the southwest of Rum.

NATURAL HISTORY

Once a millionaire's sporting estate, the remote and untamed Isle of Rum is now an important conservation and research centre for long-term ecological studies. The international importance of its diverse range of wildlife has led to it being designated a National Nature Reserve, a Biosphere Reserve, a National Scenic Area, Special Protection Area for Birds, and a Site of Special Scientific Interest. Its unique volcanic geology has led to seven Geological Conservation Review Sites on the island.

In particular, Rum is renowned for its red deer population and the scrutiny these magnificent beasts have received from scientists based at Cambridge and other universities since 1972. In 2006, the red deer rut at Kilmory was shown live on the BBC TV *Autumnwatch* programmes, with Simon King as presenter. Several herds of feral goats, probably introduced as domestic animals in the early 19th century, live among the precipitous cliffs along the west coast. The strong little Rum ponies are the only working animals on the island, and are used to carry the bodies of shot deer down from the hills to Kinloch during the culling season.

Rum is also renowned as an important breeding ground for several species of seabird, and has been designated a Special Protection Area for Birds since 1982. The most famous are the 60,000 pairs of Manx shearwater which breed in burrows on the Cuillins during the summer.

Other seabirds that breed on Rum's south coast are thousands of pairs of guillemot and kittiwake, along with razorbill and oystercatcher. The red-throated diver is also present in smaller numbers inland. Birds of prey are well represented by golden eagle, merlin, kestrel and the recently reintroduced white-tailed sea eagle.

One of the first objectives of Scottish Natural Heritage was to restore the native woodlands that once use to grow on Rum. Over the years, over a million trees have been planted around Loch Scresort and up Kinloch Glen, helping to re-establish woodland plants such as bluebell, primrose, wood-sorrel and stitchwort. Around the former township of Harris in the southwest, the grazing by Highland cattle allows wildflowers such as wild thyme to thrive, while on the mountain slopes to the east many rare alpine plants, such as northern rock-cress and arctic sandwort, still flourish.

How to Get There
By sea Caledonian MacBrayne operate a passenger ferry between Mallaig and the Small Isles. For more details contact their reservations office (tel. 08705 650000) or visit their website: www.calmac.co.uk

Day trips are also operated from Arisaig during the spring and summer by Arisaig Marine (tel. 01687 450224).

Ordnance Survey Maps
Landranger 1:50,000 series No. 39

Tourist Information
Nearest office: Mallaig Tourist Information Centre, East Bay, Mallaig PH41 4QS (tel. 01687 462170). Website: www.road-to-the-isles.org.uk

Where to Stay
There is bed and breakfast accommodation and a self-catering hostel in Kinloch Castle. For more details visit the island website: www.isleofrum.com (tel: 01687 462037). There is a campsite in Kinloch, but this is not recommended by the island's own website! Wild camping is also permitted on the island with prior permission from the reserve office (tel. 01687 462026). Caution: midges are a major problem during the summer!

Above *Located at the head of Loch Scresort, Kinloch Castle was built by Lancashire millionaire, George Bullough, at the end of the 19th century. Built of red sandstone imported from Dumfriesshire, it was one of the first buildings in Scotland to have electricity powered by its own hydro-electric generator. Today, this monument to Victorian opulence houses a hostel and bed and breakfast accommodation for visitors to Rum.*

Island Walks
With no traffic, the largest and certainly the most remote and wildest of the Small Isles, Rum can only be properly explored on foot. Well-made tracks run from Kinloch to Kilmory and its beach in the north and to Harris and the Bullough Mausoleum in the southwest. More suitable for the serious hillwalker are footpaths from Kinloch to Dibidil and the ruined lodge at Papadil in the south and down Glen Shellesder in the west. Rum has a much higher average rainfall than the other Small Isles and walkers are advised to be prepared for any eventuality.

George Bullough and his Orchestrion
At the end of the 19th century, George Bullough turned Rum into a millionaire's sporting paradise. Employing hundreds of craftsmen, Kinloch Castle took three years to build and, when complete, boasted its own electricity supply, telephone system, ballroom, heated conservatories, aquariums and an Orchestrion. The latter is an unusual electric organ that uses punched card rolls and sounds just like an orchestra. It is one of only three that exist in the world and the only one that still works.

THE SMALL ISLES: EIGG

For such a small island, Eigg and its people have had more than their fair share of trials and tribulations. From the massacre of the first early Christians and murderous inter-clan feuding when the entire population was wiped out, to wholesale emigration during the 'clearances' and a string of worse-than-useless absentee landlords, the islanders are, at long last, in charge of their own destiny. Today, environmentally-friendly farming and eco-tourism are part of the island's economic upturn.

Ferries:
1 to Rum and Mallaig
2 to Arisaig (summer only)
3 to Muck

Below *An Sgurr (or The Notch) is not only the highest point on Eigg (1,289ft) but also the largest exposed piece of pitchstone in the United Kingdom. On a clear day the view from the summit, also the site of a Iron Age hillfort, of the surrounding Hebridean islands is breathtaking.*

HISTORY

From Stone Age tools and Bronze Age burial mounds to Iron Age hillforts, Eigg is littered with archæological remains that show the island has been occupied by man for over 5,000 years. The early Christian missionaries arrived on Eigg in the early 7th century, when St Donnan arrived from Iona and established a monastery here. His presence on the island was not welcomed by the local ruler, and both he and his 52 monks were murdered in 617. The monastery soon re-established itself only to be crushed again when the pre-Christian Vikings arrived in the region at the end of the 8th century. Artefacts found on Eigg, such as jewellery and the remains of a boat, show that the Vikings probably settled on the island. Viking rule over the island lasted until 1266 when the Treaty of Perth was signed and the Western Isles came under Scottish rule.

By the early 14th century, Eigg had been granted to the MacDonalds of Clanranald, but the island's remoteness from the seat of power in Edinburgh led to much lawlessness in the region. Bitter feuding between the MacDonalds and the MacLeods of Harris in the 16th century led to the massacre of Eigg's entire population of 395 men, women and children. The site of this terrible deed, Massacre Cave (or Uamh Fhraing), can be seen on southeast coast a short distance from Galmisdale.

By the early 19th century, the fertile island of Eigg was a major producer of soda ash - made by the harvesting and drying of kelp - which, along with sheep farming and the growing of potatoes, kept the island's population of over 500 fully employed. However, this all changed when the soda ash market collapsed after the end of the Napoleonic Wars. The family fortunes of the MacDonalds also took a downward spiral and, in 1827, Eigg was sold to Dr Hugh MacPhearson.

Worse was to follow with the forcible 'clearances' of crofters from townships such as Upper and Lower Grulin in the 1850s. The island was then turned into a sporting estate for the wealthy; many islanders emigrated to Canada and by 1871, the population had dwindled to only 290. Over the next hundred years Eigg changed hands many times, until it was sold to the German artist,

Right *The volcanic geology of the southern part of Eigg is of tremendous interest to geologists. Here, overlooked by basalt sills originally formed 60 million years ago, a family of seals bask on the rocks in Kildonnan Bay. Before the creation of crofts in the early 19th century, Kildonnan was the largest and most fertile township on Eigg with just under 600 acres of arable and pasture land.*

Maruma, in 1995. By then, with little input from its often-absent owners, the island's disgruntled community had dwindled to under 70 people. Luckily, this sad state of affairs came to an end when, in 1997, Maruma put Eigg up for sale. Seizing their opportunity the islanders, along with the Scottish Wildlife Trust and the Highland Council, formed the Isle of Eigg Heritage Trust and raised the £2 million needed to buy the island. Since then, with the help of a new pier and improved transport links to the mainland, there has been a regeneration of farming, an upsurge in new businesses and tourism and, for the first time in over 100 years, an increase in the island's population.

NATURAL HISTORY

The second largest of the Small Isles, Eigg is just over 5 miles long and 3½ miles wide and, for such a small island, has a wide diversity of natural habitats ranging from bog and heather moorland to mixed woodland, farmland and a rocky coastline. A paradise, not only for the geologist attracted to its unique volcanic formations, Eigg is also a magnet for the naturalist. To delight the botanist, nearly 500 species of wild flower, including rare orchid, alpine flowers, moss and lichen, have been recorded in recent years. For those interested in birdlife, 190 species are known to regularly reside on or visit the island. The seas around Eigg are regularly visited by dolphin, porpoise and whale.

HOW TO GET THERE

By sea Caledonian MacBrayne operate a vehicle and passenger ferry between Mallaig and the Small Isles. For more details contact their reservations office (tel. 08705 650000) or visit their website: www.calmac.co.uk

Day trips are also operated from Arisaig during the spring and summer by Arisaig Marine (tel. 01687 450224).

ORDNANCE SURVEY MAPS

Landranger 1:50,000 series No. 39

TOURIST INFORMATION

Nearest office: Mallaig Tourist Information Centre, East Bay, Mallaig PH41 4QS (tel. 01687 462170). Website: www.road-to-the-isles.org.uk

WHERE TO STAY

Accommodation on Eigg comprises several guest houses, self-catering establishments, a hostel and a campsite. For more details visit the island's website: www.isleofeigg.org

ISLAND WALKS

With virtually no traffic, Eigg is ideal to explore on foot. The climb to the island's highest peak, An Sgurr (1,289ft), from which there are stunning views, is well worth the effort. Other destinations include Laig Beach and the Singing Sands, the deserted villages of Upper and Lower Grulin and the Massacre Cave.

THE SMALL ISLES: MUCK

The smallest of the Small Isles, windswept Muck and its inhabitants have, for centuries, struggled to make a living from farming and fishing. Due to the 'clearances' in the 1820s, when many islanders were evicted from their homes, and lack of investment during the 20th century, Muck's population had dwindled to just a handful by the 1970s. This trend has now been reversed with the recent upgrading of the island's infrastructure and improved links to the mainland.

ferry to Rum

Eilean nan Each

Gallanach

MUCK

Port Mòr

Beinn Airein ▲

pier

ferry to Eigg and Mallaig

0   kilometres   2

0   miles   1

Below *The old pier at Port Mor could only be used at high tide and has now been replaced by a more modern structure closer to the mouth of the bay. From a high of 321 in 1821, Muck's population is now down to about 30.*

HISTORY

Little is known of the early history of Muck, but it is likely that it has been farmed since the Middle Stone Age. There are also a few clues, in the shape of burial cairns, to occupation during the Bronze Age. Caistel nan Duin Bhan, a fortified rock at the entrance to Port Mor, probably dates back to the Iron Age. Early Christian hermits were followed by the Vikings, but little more is known of the island until the 16th century, when MacIan of Ardnamurchan leased it from the Bishop of the Isles.

By the 17th century Muck was in the hands of the MacLeans, who owned and farmed it until the mid-19th century. During the 'clearances' of the 1820s, however, most of the islanders, who had previously scratched a living from growing crops such as barley, fishing and kelp production, were evicted. Many emigrated to Cape Breton Island off the eastern coast of Canada. Changing hands in 1857, Muck eventually ended up being sold in 1896 to the MacEwan family, who still own the island today. In recent years major improvements to the island's infrastructure, such as the building of a new school, an all-weather pier, the installation of wind turbines, and the introduction of new vehicle ferry, have helped to arrest the island's declining population. Today, Muck is run as a hill farm and, together with tourism during the summer months, its economy depends almost entirely on the export of calves and lambs.

NATURAL HISTORY

Consisting mainly of volcanic basalt, the island of Muck has been farmed for hundreds of years and, until 1922, was totally treeless. Since then, however, several plantations of evergreens have been established on the island. Due to sheep grazing, Muck is not renowned for its flora but is rich in both bird and marine life. Together with the many seabirds that breed along its rocky coastline, there about 40 species of bird that normally nest on the island. Rarer visitors have also included the corncrake and Manx shearwater and, during the winter, the island is home to several hundred greylag geese. Muck's shoreline, rich in marine life carried there by the Gulf Stream, is regularly visited by Atlantic grey seal while, out to sea, porpoise and minke whale are a common sight in late summer.

HOW TO GET THERE
By sea Caledonian MacBrayne operate a vehicle and passenger ferry between Mallaig and the Small Isles. For more details contact their reservations office (tel. 08705 650000) or visit their website: www.calmac.co.uk

Day trips are also operated from Arisaig during the spring and summer by Arisaig Marine (tel. 01687 450224).

ORDNANCE SURVEY MAPS
Landranger 1:50,000 series No. 39

TOURIST INFORMATION
Nearest office: Mallaig Tourist Information Centre, East Bay, Mallaig PH41 4QS (tel. 01687 462170). Website: www.road-to-the-isles.org.uk

WHERE TO STAY
There are four self-catering properties, a self-catering bunkhouse, a bed and breakfast establishment and a small hotel on Muck. Camping is allowed on the island. For more details

contact Mallaig Tourist Information Centre (see above) or visit their website: www.road-to-the-isles.org.uk/muck.html

ISLAND WALKS
Visitors can walk anywhere on the island apart from in private gardens. The best way to see the island is to walk from the pier at Port Mor along the only road on the island to Gallanach Bay. From here, it is an easy climb to the island's highest point, Ben Airean.

Above *The island graveyard at the head of Port Mor contains this grave of an unknown rating who was killed when his ship,* HMS Curacao, *was accidently rammed by* RMS Queen Mary *to the southwest of Tiree in 1942. Behind the graveyard are the ruins of the village of Keil, which became derelict after the 'clearances' of 1828 when half the population of the island was evicted.*

COLL

D uring the 19th century, following 500 years of rule by the paternalistic chiefs of the MacLeans of Coll, many of the people of this magical island were forcibly evicted from their crofts and given one-way tickets to North America, Australia and New Zealand to make way for large-scale dairy farming. More recently, improved communications and the growth in eco-tourism have brought greater stability to the island's tiny population of around 170.

HISTORY

Once home to early Neolithic farmers and their standing stones; Bronze Age settlers and his crannogs (man-made islands on inland lochs); and Viking settlers, Coll became part of the Norse Kingdom in the 9th century, until the Treaty of Perth in 1266. In the 14th century, the island was granted by the Lord of the Isles to the first chief of the MacLeans of Coll. To protect their island home, the MacLeans built their first castle at Breachacha in the south of the island during the 15th century. In the late 16th century, the nearby stream which runs into Loch Breachacha was the site of a bloody battle between the MacLeans of Coll and the MacLeans of Duart when the latter invaders were severely beaten.

For 500 years, successive chiefs of the MacLeans ruled Coll in a paternalistic manner, seeking to improve both their lands and the lives of their tenants. By 1841, however, fuelled by failed potato harvests and an unsustainable population of over 1,400, both the MacLeans and their tenants had fallen on hard times and, in 1856, most of the island was sold to John Lorne Stewart, then the Duke of Argyll's chancellor.

Over the next few years Stewart cleared many of the crofts, with hundreds of people being forcibly evicted and given one-way tickets to either North America, Australia or New Zealand; in their place he introduced dairy farming with imported cattle from the mainland. Even this bold 'experiment' was not a total success and, by the early 20th century, much of the unproductive land on the island had returned to nature and the population continued to dwindle. Despite the introduction of new, but short-lived, industries such as the growing of tulip and daffodil bulbs and the manufacture of perfume, by the 1960s the population had dropped below 150. Since then, mainly thanks to improved communications and a growing tourist industry, the population has stabilised at around 170 and Coll's prospects are looking much brighter.

NATURAL HISTORY

A wildlife paradise, Coll is famed for both its rare wild flowers and birdlife. In particular, the large nature reserve owned by the Royal Society for the Protection of Birds in the southwest of the island, which includes habitats ranging from shell-sand beaches, sand dunes and machair (the distinctive type of coastal grassland found in the Hebrides) to farmland and moorland, is the breeding ground of one of Britain's rarest birds - the corncrake. Once common throughout Britain, its numbers have been dramatically reduced by modern farming methods and is now only found on a few of the Hebridean islands. Since 1991, when the RSPB bought the reserve, careful management of the land

Above *Arinagour is the main settlement on Coll. Its tiny old harbour is overlooked by the award-winning and popular Coll Hotel.*

Below *Visited by Dr Samuel Johnson and James Boswell on their epic trip to the Hebrides in 1773, the 'new' Breachacha Castle was built in 1750 by the 12th chief of the MacLeans. After many years of neglect, its new owners are currently restoring it to its former glory. Nearby is the 'old' Breachacha Castle which was built in the 15th century as a MacLean stronghold and is also now a private residence. Until World War I, the castle's farms produced the famous Coll cheese which found popularity at the Houses of Parliament in London.*

has seen a big increase in the corncrake population. The reserve is also home to large numbers of wading birds, flocks of over-wintering barnacle and white-fronted geese, two rare bees – the great yellow bumblebee and a mining bee – and about 300 species of flowering plants, including rare orchids, which grow in profusion on the lime-rich machair. Otters are frequent visitors along the coastline, while out to sea it is quite common to see seal, dolphin, basking shark and whale. For more information about the reserve contact the RSPB information centre at Totranold (tel. 01879 230301) or visit their website: www.rspb.org.uk/scotland

How to Get There

By sea Caledonian MacBrayne operate a vehicle and passenger ferry between Oban, Coll and Tiree. On Thursdays during the summer, the service extends to Castlebay on the island of Barra. For more details contact their reservations office (tel. 08705 650000) or visit their website: www.calmac.co.uk

By air A newly constructed airstrip on Coll eagerly awaits the much-heralded service from Oban Airport. British Airways operate regular daily flights from Glasgow to the neighbouring island of Tiree (see p.104).

Ordnance Survey Maps

Landranger 1:50,000 series No. 46

Tourist Information

Nearest office: Oban Tourist Information Centre, Argyll Square, Oban, Argyll (tel. 01631 563122).
Website: www.oban.org.uk
Alternatively visit the island's two websites: www.visitcoll.co.uk or www.isleofcoll.org

Where to Stay

There is limited choice of accommodation on Coll, ranging from the popular Coll Hotel and a few self catering and bed and breakfast establishments to a static caravan and camping site. For more details contact Oban Tourist Information Centre or the island's two websites (see above).

Island Walks

Only 13 miles long and 3 miles wide and with over 20 secluded sandy beaches, the island of Coll is perfect for a walking holiday. The RSPB organise weekly guided walks in the summer from their information centre at Totranold. (tel. 01879 230301). On fine days there are good views of the surrounding islands from Coll's highest point, Ben Hogh. Cycle hire is also available in Arinagour.

Below *In the southwest of Coll, Feall Bay is one of the many fine shell-sand beaches found along the coastline. Located within the RSPB nature reserve, the crescent-shaped beach is backed by massive sand dunes and acres of lime-rich machair, where nesting lapwings are a common sight in late spring.*

TOTRONALD STANDING STONES

About 4,500 years old and located close to the RSPB information centre at Totronald, two solitary standing stones are the oldest recorded work of Neolithic man found on Coll. Called Na Sgeulachan (or 'teller of tales'), their significance to these early farmers is now lost in the mists of time.

TIREE

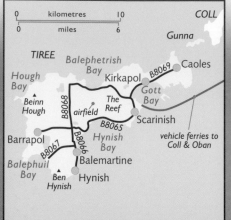

Owned for over 300 years by the Dukes of Argyll, Tiree has seen its population drop from nearly 4,500 in 1831 to just under 800 today. However, those islanders that still remain can, in the summer, enjoy its long hours of sunshine, miles of white sandy beaches, turquoise seas, acres of closely cropped machair and some of the best birdlife in the United Kingdom. In the winter, they batten down the hatches to prepare for the storms that blow in from the Atlantic.

HISTORY

Although it is likely that Tiree has been inhabited for over 7,000 years, the earliest remains found on the island date from the Bronze Age which lasted between 2500BC and 600BC. Probably the most famous religious artefact from this period is the Ringing Stone with its cup-shaped markings, which is located on Tiree's north coast just over one mile east of Balephetrish Bay. The later Iron Age is well represented on Tiree with over 20 forts, duns and brochs scattered around its coastline. The best preserved of these, Dun Mor Vaul, can be found a short distance up the north coast from the Ringing Stone. By the 6th century AD, Tiree was being visited by the early Celtic Christians from Ireland who founded a monastery, known as Campus Luinge, where wayward monks were sent to pay penance for their crimes. Historians still argue over the possible location of this monastery, some believing it was at Kirkapol while others

TIREE AND THE WEATHER

Tiree is one of the sunniest and windiest places in the United Kingdom. Detailed weather records have been kept on the island since 1926, when a local headmaster first set up a weather station in the grounds of his school at Cornaig. Tiree held the sunshine record for the UK in 2006, with 16½ hours on 26 June. High winds are common and in winter this treeless island is regularly battered by Atlantic gales with windspeeds sometimes reaching nearly 120mph. Tiree's winds make it one of the most midge-free locations in Scotland during the summer!

Right *During World War II Tiree was an important RAF base for anti-submarine and weather flights over the Atlantic. On 16 August 1944 two Halifaxes collided in bad weather over Tiree killing all of those on board, including Czech, Australian, Canadian and British airmen. Their neatly kept graves can be seen today in Soroby cemetery in the southwest of the island.*

think that it was on the site of Soroby graveyard. By the early 9th century Tiree, along with the rest of the region, was beginning to feel the full impact of Viking invaders. It is likely that many of these Norsemen settled on the island, which remained under Norwegian control until the signing of the Treaty of Perth in 1266, when the Western Isles became part of the kingdom of Scotland.

There then followed 400 years of inter-clan feuding over the ownership of Tiree, with the worst excess taking place in 1543 when the Campbells of Argyll attacked the then owners, the MacLeans of Duart, and laid waste to Tiree. The MacLeans struggled on for another century, but by then they were bankrupt and, seizing his chance, the 8th Earl of Argyll took control of the island in 1679. Over 300 years later, most of Tiree is still owned by the Duke of Argyll.

Introduced in the mid-18th century, the kelp industry on Tiree grew to become one of the main sources of income for islanders. Ash from burnt kelp, rich in soda and potash, was an important ingredient in the manufacture of glass and soap during the 18th and 19th centuries. Due to cheaper competition from Europe at the end of the Napoleonic Wars, the bottom fell out of the local soda ash market and by the 1830s production had ceased completely. Still an important fertiliser for farmers,

Below *In the far southwest of Tiree, its clean white sand washed by Atlantic rollers, Balephuil Bay is one of many fine beaches on the island. Not only is Tiree one of the sunniest places in Britain but it is also the windiest - which helps to explain why the PWA Windsurfing World Cup is held here every October.*

SOROBY AND THE McLEANS

In the southwest of Tiree, Soroby graveyard is possibly the site (the other is at Kirkapol) of Campus Luinge, a monastery founded in 565AD by Baithenes, one of St Columba's followers. The monastery was destroyed by the Vikings, but the graveyard contains the graves of the Clan McLean chiefs who ruled Tiree from 1390 to 1679. This well-preserved ancient decorative cross (right), which is inscribed on one side in Latin and on the other in Celtic, is known as MacClean's Cross.

Above *Washed up forever on the sands of Scarinish Harbour are all that remains of the double-topsail schooner* Mary Stewart. *She was built by Barclay of Ardrossan in 1868 and, after a stint around the Irish coast, ended her days trading up and down the west coast of Scotland until the 1930s.*

Below *The lime-rich machair, home to the rare mining bee, in all of its glory near Balevulin in the northwest of the island. One of the most fertile of the Hebridean islands, Tiree is famous for its annual agricultural show where the island's highly sought after sheep and cattle are on show.*

kelp is once again collected on Tiree, this time for use as alginates for food thickening.

By the 19th century, the writing was also on the wall for many of Tiree's tenant farmers. Whole townships were cleared to make way for large sheep farms and, with the population peaking at 4,450 in 1831, there were hard times ahead. By the mid-19th century, with failing potato harvests and an unsustainable population, hundreds of people left Tiree for Canada. But worse was to follow for those left behind and by 1864, the year of the potato blight, the impoverished islanders were depending on charitable aid to stave off starvation.

Following years of unrest over unfair rents and lack of security of tenure, both on Tiree and on the mainland, land reforms were finally introduced in the Crofters' Holdings (Scotland) Act of 1886. However, despite a much fairer distribution of land on the island, Tiree's population continued to decline for the next 100 years. By 1991 it had bottomed out at 768 and has now stabilised at around that figure – with 25% of the population now over 65, nearly twice the national average!

Tiree's dwindling fortunes were given a boost during World War II when a RAF airfield was built in the centre of the island. Halifaxes and Hudsons, flown by airmen from many Allied nations, patrolled the Atlantic in all weathers in search of German U-boats and to collect valuable weather data. As a result, by the end of the war, Tiree had an excellent network of roads and a long runway.

Today, regular flights from Glasgow, a modern ro-ro ferry link with Oban and hi-tech communications all help contribute to the island's future success. While crofting, sheep and cattle raising are still important economic activities on this fertile island, tourism is today playing an increasingly vital role.

NATURAL HISTORY

A mecca for birdwatchers, Tiree is internationally renowned for its visiting corncrake population which, thanks to the restoration of traditional farming practices and land management, has risen during the summer months from about 100 calling males in 1988 to just over 300 by 2006. This is by far the largest single population of corncrakes in the United Kingdom. In addition to the uncut grass cover required for this most endangered bird, Tiree has a wide range of other habitats that support many other species of bird. Large numbers of barnacle and Greenland white-fronted geese are regular winter visitors while the greylag is a permanent resident. In the summer the machair, ablaze with

STAFFA AND FINGAL'S CAVE

The island of Staffa, its world-famous Fingal's Cave, and the nearby Treshnish Isles can be reached either from Fionnphort or Ulva Ferry on Mull or from Scarinish Harbour on Tiree. Landings on the island depend on sea conditions. Staffa was given to The National Trust for Scotland by an American, John Elliott Jnr, in honour of his wife Eleanor Thomas Elliott in 1986. Popular since Victorian times, Staffa's fantastic and regularly shaped basalt columns and caves attract thousands of visitors a year. Among its many famous visitors was the composer Felix Mendelssohn who wrote his Hebrides Overture after a visit to the caves in 1829.

Right *In 1834, following years of shipwrecks on reefs 11 miles southwest of Tiree, Alan Stevenson, an uncle of Robert Louis Stevenson, was appointed to build the 138ft-tall Skerryvore lighthouse. The granite used in its construction was prefabricated at Hynish in the southwest of Tiree. Completed in 1842, the lighthouse was considered one of the engineering wonders of the world. The dock, workshop, lodgings and signalling station at Hynish (right) that were used in the construction and maintenance of the lighthouse have all been renovated by the Hebridean Trust.*

wildflowers and the sound of skylarks, and the shallow lochs are alive with waders such as snipe, oystercatcher, lapwing and dunlin. Loch Bhasapol, in the northwest of the island, has a bird hide overlooking the loch's western shore from where such rarities as ring-necked duck and Slavonian grebe have been seen in recent times. During the spring and autumn, Tiree is regularly visited by rare migratory birds, some blown thousands of miles off course across the Atlantic from North America.

Introduced to Tiree by the Duke of Argyll's estate in 1827 the brown hare, which is now quite rare on the mainland, is still a relatively common sight on the island. It is reckoned that Tiree supports a population of around 500 brown hares, the fastest wild animal in Britain.

HOW TO GET THERE
By sea Caledonian MacBrayne operate a vehicle and passenger ferry between Oban, Tiree and Coll. On Thursdays during the summer, the service extends to Castlebay on the island of Barra. For more details contact their reservations office (tel. 08705 650000) or visit their website: www.calmac.co.uk
By air British Airways operate regular daily flights from Glasgow to Tiree. For more details contact BA (tel. 0870 850 9850) or visit their website: www.ba.com

ORDNANCE SURVEY MAPS
Landranger 1:50,000 series No. 46

TOURIST INFORMATION
Nearest office: Oban Tourist Information Centre, Argyll Square, Oban, Argyll

(tel. 01631 563122) or visit the island's community website: www.isleoftiree.com

WHERE TO STAY
There is a reasonable range of accommodation available on Tiree ranging from two hotels and bed and breakfast and self-catering establishments to a hostel and campsite. For more details contact Oban Tourist Information Centre or visit the island's website (see above).

ISLAND WALKS
Being a very flat island with only two hills of any significance, Tiree lends itself to a relaxing walking (or cycling) holiday. There are many pleasant walks to be taken along the island's many fine, human-free, sandy beaches and the adjoining well-manicured machair.

KIRKAPOL CHAPELS

While Soroby graveyard (see p.105) is usually accepted as the site of Campus Luinge, a monastery founded in 565AD by St Baithene, where wayward monks paid penance for their crimes, there is another school of thought that believes the site is at Kirkapol in the east of Tiree. Here, overlooking the broad expanse of Gott Bay, are two ruined chapels – the smaller dating from 13th century and the other from the 14th century. The larger graveyard is dedicated to St Oran, who was a relative and follower of St Columba.

ISLE OF MULL

The fourth largest of the Scottish islands, Mull and its people were in the front line of clan and religious warfare during the 17th century. Becoming a major victim of forcible evictions during the notorious 19th century 'clearances', the island's population has seen a dramatic decline until recent years. Today, attracted by the islands stunning natural beauty, wildlife and rich historical heritage, hundreds of thousands of tourists throw an economic lifeline to Mull's ageing population.

Below *Built as a fishing port in the late 18th century, colourful Tobermory is now the main town on Mull. A galleon from the defeated Spanish Armada of 1588, loaded with a vast amount of gold coins, is supposed to have sunk in Tobermory Bay when taking on provisions. The wreck, with its hoard of treasure, has never been found. Tobermory is also the setting for the long-running BBC TV children's series,* Balamory, *in which local children have been cast as extras. For more information visit the town's website: www.tobermory.co.uk*

HISTORY

Mull has been inhabited since prehistoric times and the island has several important archæological sites that are over 4,000 years old. The best preserved are the stone circle at Lochbuie on the southeast coast and several standing stones just to the east of Dervaig in the north, at Gruline in the west and at Uisken in the southwest. The remains of Bronze Age burial cairns can be found near Gruline and near the main coast road, about four miles north of Salen. The later Iron Age is well represented by the ruins of many forts or duns. Of these, the best coastal examples are Dun-nan-Gal, two miles northwest of Ulva Ferry, Dun Aisgain and Dun Ban, both on the north shore of Loch Tuath and two on the remote western coast at Port Haunn. Dun Urgadul is a good inland example just west of Tobermory.

The first of the early Celtic Christians, St Columba, and his band of followers arrived on the neighbouring island of Iona (see pages 114–115) from Ireland in 563AD. From his island base, St Columba spread the Christian word far and wide across Scotland and northern England. From the end of the 8th century, both Mull and little Iona bore the brunt of attacks from Viking raiders until the whole region finally came under Norse control, after King Magnus III (aka Barelegs) of Norway took over the Orkneys, the Hebrides and Isle of Man by force at the end of the 11th century.

Norse rule ended after the signing of the Treaty of Perth in 1266 and the Isle of Mull came under the control of the Lords of the Isles. For three centuries, their stronghold on Mull was at Aros Castle, now a ruin, located on the east coast just north of Salen. The two dominant clans on Mull were the MacLeans, based at Duart Castle, and the MacKinnons, who were based on the Isle of Skye but who also owned Iona and lands on Mull. By the 16th century, however, the Campbells, led by the Protestant Earl of Argyll, had taken over as Lord of the Isles.

By the late 17th century they had taken over Mull, including Duart Castle and the MacLean lands, by force. Aros Castle became their main stronghold while Duart Castle was garrisoned by Government troops until 1750.

While the islanders had suffered greatly from the forcible transfer of power on Mull, within 100 years their lot had become much improved when the 5th Duke of Argyll, previously the MP for Glasgow and later, for Dover, introduced fairer tenancy agreements throughout his Scottish estates. Improvements were also made to the island's infrastructure, including the building by the newly founded British Fisheries Society of a new fishing port at Tobermory. By 1793, the new village consisted of two streets of houses, a store house,

customs house and harbour, and had attracted 32 settlers and their families. Tobermory's prosperity as a port grew further when the Caledonian Canal was opened in 1822.

By this time Mull's population had increased to over 10,000, a figure which had become unsustainable for the island's resources. A major decline in the kelp burning industry, and failure of successive potato crops, hastened mass emigration to the New World. Not everyone left peacefully, and there were many terrible examples of forced evictions of tenants and their families from their homes. By 1881 Mull's population had halved to 5,229, and continued to decline until it reached a low of just over 2,000 in 1971.

Today, thanks to improved communications with the mainland and the growth of island industries such as forestry, agriculture, fish farming and tourism, the population has once again started to increase.

Above *A well known landmark to passengers on the Calmac ferry from Oban to Craignure, Duart Castle has been the ancestral home of the MacLeans for around 750 years. Considerably rebuilt in 1673 by Sir Allan MacLean, the castle was surrendered, along with all of the MacLean lands on Mull, to the Duke of Argyll in 1691. It was then used as a garrison for Government troops until 1750 before falling into disrepair. The castle was bought by Sir Fitzroy MacLean in 1910 and subsequently restored to its current condition. The castle is open to the public. For more details visit the castle's website: www.duartcastle.com*

However, the island's rugged natural beauty, rich tapestry of wildlife and historical heritage have together made tourism the most important economic activity on Mull. Each year, mainly between June and September, over half a million visitors, over 100,000 cars and thousands of coaches – many on their way to the island of Iona – descend on the island, clogging its single-track (with passing places!) road system. However, many islanders, a high percentage of whom are aged over 60, depend on this mass influx for their livelihood.

NATURAL HISTORY

The fourth largest of the Scottish Islands with a coastline of around 300 miles, Mull has a wide range of natural habitats ranging from mountain peaks, open moorland and large tracts of mainly coniferous forest to freshwater lochs, sea lochs, marshland and sandy beaches.

A paradise for birdwatchers, Mull is home to around 250 species of bird and a stopping-off point for many migrants during spring and autumn. In particular, the mountains and moorland areas are home to many birds of prey including hen harrier, kestrel, sparrowhawk, peregrine falcon and merlin. Mull has one of the largest concentrations of golden eagle in the world while the rare white-tailed sea eagle, with its 9ft wingspan, is also on the increase in the island. Ptarmigan are seen on the higher slopes. Mull also has a large population of owls including barn owl, long-eared owl, tawny owl and visiting short-eared owl.

Below *Many islanders emigrated from Mull to Canada during the 19th century. To remind the emigrés of their lost home, the Canadian city of Calgary is named after this beautiful bay on Mull's west coast. Calgary Bay, with its gently shelving white sandy beach and crystal clear waters, is also home to breeding sand martins between spring and autumn. Droughts in their African wintering grounds have severely reduced their population in recent years.*

Right *Overlooked by Mull's highest peak, Ben More (3,169ft), the little island of Eorsa rises out of the waters of Loch na Keal on Mull's west coast. Used for sheep grazing, this privately-owned island has never been inhabited.*

Mull's long coastline attracts many waders and migrants including greenshank, redshank, snipe, whooper swan and whimbrel. Teal, widgeon, shelduck, black-throated and great northern diver are visitors to the sea lochs in winter, while the inland freshwater lochs attract the red-throated diver in spring and summer. The island is regularly visited by thousands of migatory birds during spring and autumn. The list of rare visitors in recent times is impressive and includes lesser yellow leg, nightjar, pectoral sandpiper, bee-eater and an American golden plover.

For more details about the birds of Mull visit the website: www.mullbirds.com

Mull is also home to many mammals including the shy otter which, with patience, can be seen along the shores of the sea lochs. Others regularly seen on Mull include polecat, stoat, weasel, mountain hare, the destructive mink – escapees from mink farms – and large herds of red deer.

For wildlife expeditions contact Isle of Mull Wildlife Expeditions (tel. 01688 500121) or visit their website: www.torrbuan.com

Mull is also rich in plantlife and, during the early summer, many parts of the island are ablaze with the colour of wild orchids including northern marsh, heath spotted, early purple, lesser and greater butterfly and fragrant orchids.

With over 5,000 recorded species of flora, Mull is a haven for botanists.

For an opportunity to see Mull's flora contact Discover Mull Landrover Tours (tel. 01688 400415) or visit their website: www.discovermull.co.uk

Many species of butterfly have also been recorded on the island, including common blue, peacock, painted lady, Scotch argus and marsh fritillary.

The waters around Mull's long coastline are a favourite location for

TRESHNISH ISLES

Uninhabited since the early 19th century, the Treshnish Isles are a group of islands located about four miles from the northwest coast of Mull. Unusually shaped due to their volcanic formation millions of years ago, they appear like a line of battleships on the horizon. The most distinctive, Bac Mòr, is also appropriately known as Dutchman's Cap. Famed for their unique geological formations, colonies of seal, seabird, overwintering wildfowl, archaeological sites and, unusually, a population of house mice, the Treshnish Isles have been owned by the Hebridean Trust since 2000 and are designated as a Site of Special Scientific Interest and a Special Protection Area for Birds. For more information visit the Hebridean Trust website: www.hebrideantrust.org Boat trips to the islands operate from both Mull and Tiree.

whale, dolphin, porpoise and basking shark watching. The Hebridean Whale and Dolphin Trust, which is based on the island, has pioneered local conservation and awareness of these amazing creatures. For more information about their work visit their website: www.whaledolphintrust.co.uk

For whale watching trips contact Sea Life Surveys (tel. 01688 302916) or visit their website: www.whalewatchingtrips.co.uk

HOW TO GET THERE
By sea Caledonian MacBrayne operate frequent vehicle and passenger ferries between Oban and Craignure (Mull) and between Lochaline and Fishnish (Mull). A vehicle and passenger ferry also operates between the Isle of Iona and Fionnphort (Mull). For more details of all these services contact the Caledonian MacBrayne reservations office (tel. 08705 650000) or visit their website: www.calmac.co.uk

ORDNANCE SURVEY MAPS
Landranger 1:50,000 series Nos. 47, 48 & 49

TOURIST INFORMATION
Nearest offices: Craignure Tourist Information Centre, The Pier, Craignure, Isle of Mull PA65 6AY (tel. 08707 200 610); Tobermory Tourist Information Centre, Main Street, Tobermory, Isle of Mull PA75 6NU

ULVA AND GOMETRA

Linked by a short bridge, these two privately-owned islands (6½ miles long x 3 miles wide) are located a short distance from the west coast of Mull across the Sound of Ulva. In recent years, archæologists have found traces of human occupation in a cave dating back to the early Stone Age, around 7,500 years ago and, along with several standing stones, Iron Age and Viking forts and deserted townships, both islands are rich in many other sites of historical interest. With their wide range of natural habitats ranging from mixed woodland, rocky shores, moorland and high peaks (Beinn Chreagach is the highest at 1,026ft), the islands are famed for their wide variety of plant and animal life. A major victim of the infamous mid-19th century 'clearances', Ulva and Gometra's population has fallen from a high of around 650 in 1841 to a current low of 16. Today, the main economic activities are centred around sheep and cattle farming, oyster farming and fish farming. Visitors are also welcome to enjoy the islands' natural beauty and access is via a passenger ferry from Ulva Ferry on Mull. For more information visit the islands' website: www.ulva.mull.com

Above *With no tarmac roads or traffic but miles of quiet tracks, Ulva is an accessible island ideal for the walker or mountain biker. Facilities include a red telephone box, licensed tea rooms (above) and thatched visitor centre.*

(tel. 08707 200625) or visit the website: www.visitscottishheartlands.com

WHERE TO STAY

There is a wide range of accommodation available on Mull. For more information contact Craignure or Tobermory Tourist Information Centres or website (see above), or visit the island's three websites: www.explore-isle-of-mull.co.uk www.holidaymull.co.uk www.mull.zynet.co.uk

ISLAND WALKS

From forest trails and coastal walks to hillwalking and mountain climbing, Mull has everything for the dedicated walker.

The highest point on Mull and one of only two Scottish islands (the other is Skye) to posses a Munro (ie a mountain over 3,000ft), Ben More (3,169ft) is the toughest challenge on the island. Only experienced walkers should tackle the climb to the summit from where, on a clear day, there are stunning views across to the other Hebridean islands.

For coastal walks Carsaig Bay, in the southwest of Mull, is a good starting point for walks either east or west. Highlights of these two walks include caves, basalt arches, sea stacks, waterfalls and prehistoric sites.

With around 40,000 acres of managed woodland on the island, there are many opportunities for forest walks including three waymarked walks through Forestry Commission land. Two start at or near Tobermory and head north to Ardmore Bay or south to Aros Park. A shorter trail starts at Gartmony Point, four miles north of Craignure, and ends at Fishnish. Mull's two tourist information centres have leaflets and books covering many other walks on the island. Alternatively visit the website: www.scotland.org/Walks_on_Mull

Above *Balmeanach is a small farming community located on Mull's west coast at the foot of Ben More. Out to sea are the islands of Staffa and Little Colonsay, while behind them on the horizon are the unusually shaped Treshnish Isles, now owned by the Hebridean Trust.*

Below *High and dry on Mull! Overlooked by Mull's highest peak, Ben More (3,169ft), this redundant fishing boat spends its last years quietly rotting on saltmarshes on the south shore of Loch Scridain.*

INCH KENNETH

Located close to Mull's west coast near Balnahard, the small, privately-owned island of Inch Kenneth is named after the early Celtic Christian monk St Kenneth who founded a monastery here in the 6th century AD. The island was visited by Dr Johnson and James Boswell while on their grand Hebridean tour in 1773. Today, the main buildings on the island are the ruins of a 13th century chapel together with many carved gravestones and a 19th century mansion. Inch Kenneth was once owned by Lord Redesdale, the father of the Nazi sympathiser Unity Mitford, who attempted to commit suicide at the outbreak of World War II. Seriously injured with a bullet lodged in her brain, she went to live on the island with her mother but died from her self-inflicted wounds in 1948.

There is no public access to Inch Kenneth but it can be seen clearly from the coast road on Mull's west coast.

IONA

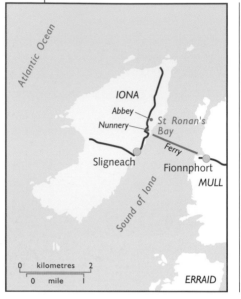

Perched in the Atlantic Ocean just a short distance from the Ross of Mull, the tiny, windswept island of Iona is famed worldwide as one of the earliest centres of Christian learning and culture. From its early days in the 6th century, its missionaries travelled far and wide to spread the Christian word. Despite destruction by the Vikings and falling into disuse after the Reformation, its famous Abbey now stands as a beacon of hope to thousands of visitors each year

HISTORY

Although Iona has probably been inhabited for at least 5,000 years, the earliest archæological remains, the Celtic hill fort on Dun Bhuirg on the island's west coast, are around 2,000 years old.

However, Iona is famed worldwide as one of the earliest centres of Christian learning in Europe. In 563 an Irish missionary, Columba, and his 12 monks landed on Iona after undertaking a hazardous sea voyage from northern Ireland. Founding a monastery on the island, Columba soon set about converting the local pagan Picts to Christianity. Soon, missionaries from Iona were travelling across Scotland and northern England to spread the word. Not only becoming a place of pilgrimage, Iona also became a centre for learning and culture, with its monks producing some of the world's earliest illuminated manuscripts, including the famous *Book of Kells*. By the end of the 8th century, however, the monks' peaceful existence on Iona was coming to an end.

Between 795 and 806 the Vikings raided Iona on three occasions, and in their wake they left death and destruction, during which the monastery was destroyed and many monks killed. Finally, the Abbot and most of his remaining monks left for relative safety in Ireland, taking with them what treasures remained.

Peace only finally returned to the region in the mid-12th century when the local chieftain, Reginald, gave permission to Benedictine monks to build a monastery again on Iona. Reginald's sister also became the first Abbess of the Nunnery of St Mary, which was built on Iona at the same time. By the end of the 15th century, the Abbey had passed into the hands of the Bishops of the Isles and was upgraded to a cathedral. Both the Abbey and the Nunnery fell into disuse during the Scottish Reformation of the late 16th century. The Nunnery is still in ruins, but the Abbey was rebuilt as recently as 1910.

By the late 17th century, Iona had become the property of the Earls of Argyll and it remained so until 1979, when it was sold to the Hugh Fraser Foundation. They, in turn, gave the island to the National Trust for Scotland.

Above *During the peak season, the main settlement on Iona, Baile Mor, sees a massive influx of tourists who take the short ferry ride from Fionnphort on Mull. All of the island's main attractions - the Nunnery, St Ronan's Church, St Oran's Chapel and the Abbey - are within a short distance of the jetty.*

Below *Sheltered from Atlantic storms in the winter, Iona's small community is strung out along the island's east coast. Its famous Abbey can be seen nestling beneath the island's highest point, Dun I (328ft). Iona was given to the National Trust for Scotland by its owner, Sir Hugh Fraser, in 1979.*

NATURAL HISTORY
For such a small island (3½ miles x 1½ miles), Iona has a surprisingly wide variety of both plant and birdlife. From over 400 species of wild flower to the elusive corncrake, there is much to discover on land, while in the waters around the island whale, basking shark, dolphin, porpoise and seal are all regular visitors

HOW TO GET THERE
By sea Caledonian MacBrayne operate a vehicle and passenger ferry between Fionnphort on Mull to Iona. For more details contact their reservations office (tel. 08705 650000) or visit their website: www.calmac.co.uk

ORDNANCE SURVEY MAPS
Landranger 1:50,000 series No. 48

TOURIST INFORMATION
Nearest office: Craignure Tourist Information Centre, the Pier, Craignure, Strathclyde PA65 6AY (tel. 08707 200610)

Above *Iona Abbey was originally built by Benedictine monks in the 12th century. It fell out of use during the Scottish Reformation in the 16th century and by the end of the 19th century had become a roofless ruin. Work on restoring it to its former glory was completed in 1910 by the Church of Scotland. A nearby graveyard not only contains the graves of many Scottish kings and chieftains, but also that of John Smith, Leader of the Labour Party until his untimely death in 1994.*

WHERE TO STAY
There is a wide selection of accommodation on Iona, ranging from hotels and bed and breakfast and self catering establishments, to a hostel and several retreats. For more details contact Craignure Tourist Information Centre (see above) or visit the island's website: www.isle-of-iona.com

ISLAND WALK
With neglible traffic and only 2½ miles of single-track roads, it is quite possible to stroll around Iona in a single day.

Above *Set in the centre of the Abbey cloister, this sculpture of the holy dove descending from heaven was created by the Russian-born sculptor, Jacques Lipchitz (1891-1973).*

KERRERA

From the sudden death of a 13th century Scottish king to the destruction of its castle and inhabitants, Kerrera's history is shrouded in the mists of time. Only 4½ miles long and 1½ miles wide, this fertile island and its tiny population are only a stone's throw from the bustling town of Oban. Today, the island's tranquility, natural beauty and wildlife make it a popular destination for active visitors wishing to escape from the hustle and bustle of modern life.

HISTORY

Apart from evidence in the form of a Bronze Age burial mound containing some pottery, very little is known of the early history of Kerrera. One of the earliest written accounts dates from 1249, when King Alexander II of Scotland massed a large fleet of warships in Horse Shoe Bay in his bid to take back territories seized by the Norwegian king, Haakon. Alexander was not a well man and he sadly died soon after setting foot on Kerrera. Since at least the 13th century, the island has been owned by the MacDougalls, who built Gylen Castle as their fortified home in the south of the island in 1587. The castle and its inhabitants were wiped out during the bitter Wars of the Covenant in 1647, when it was attacked by the Duke of Argyll's troops; but the famous Brooch of Lorne, looted from the castle at this time, was finally returned to the MacDougalls in the early 19th century.

Farming has always been important on the fertile island of Kerrera. Sheep and cattle raising are still the main economic activities but, in the past. the island's tenant farmers also grew oats, barley, potatoes and flax to supplement their incomes. Apart from farming, other island activities included kelp burning and a thriving trade in lobsters.

NATURAL HISTORY

With a human population of around 35, this totally unspoilt island is a wildlife haven. There is a well-established seal colony on the small islands just off Kerrera's northwest coast, while out to sea porpoise, dolphin and minke whale are regular visitors. With patience, otters can also be seen along the island's rocky and indented coastline. Inland, rubbing

Above *The coastal track along Kerrera's east coast passes these former crofter's cottages at Little Horse Shoe Bay. It was close to here that the Scottish king, Alexander II, died in 1249. Since the mid-19th century, Kerrera has seen its population drop from just under 200 to the present-day level of 35.*

Above *Located in the south of the island, Gylen Castle was built in 1587 as a stronghold for the MacDougalls. The castle was besieged in 1647 by the Duke of Argyll's troops under the leadership of Major-General Leslie during the Wars of the Covenant. At the end of the siege, all of the castle's inhabitants were massacred by Leslie's troops.*

shoulders with Highland cattle and sheep, is a herd of wild goat. The island also boasts a profusion of birdlife, including the majestic golden eagle, hen harrier and smaller birds such as skylark, willow warbler, whitethroat, linnet, reed bunting and twite.

HOW TO GET THERE
By sea A small passenger ferry operates to the island from the Kerrera Ferry car park, two miles southwest of Oban (for details of times tel. 01631 563665).

ORDNANCE SURVEY MAPS
Landranger 1:50,000 series No. 49

TOURIST INFORMATION
Nearest office: Oban Tourist Information Centre, Argyll Square, Oban, Argyll (tel. 01631 563122).
Website: www.oban.org.uk

WHERE TO STAY
Basic accommodation is available at two locations on Kerrera. In the south is the Bunkhouse at Lower Gylan. For more details contact the Kerrera Bunkhouse (tel. 01631 570223), website: www.kerrerabunkhouse.co.uk
In the north of the island is the

Ardantrive Hostel (tel. 01631 567180).
There is also a wide range of accommodation in nearby Oban. For more details contact the Oban Tourist Information Centre (see above).

ISLAND WALKS
With virtually no traffic or tarmac, by far the best way to explore Kerrera is on foot. From the ferry landing, a 6-mile circular walk southwards takes in Horse Shoe Bay, Gylen Castle, Lower Gylem Tearooms, Ardmore and then up the west side of the island, with its raised beaches, back to the starting point. There is also a track to Ardantrive and the Hutcheson Memorial in the north of the island.

Below *The western sea approaches to Oban are dominated by this obelisk on the northernmost tip of Kerrera. To the left is the sheltered anchorage and boatyard of Ardantrive Bay. The obelisk was erected at the end of the 19th century in memory of David Hutcheson, one of the founders of what was to become the Caledonian MacBrayne shipping company. Born in 1799, Hutcheson died in Glasgow in 1880 and was buried in the Pennyfuir Cemetery midway between Oban and Dunbeg.*

LISMORE

ismore, or the 'Great Garden' as it translates from Gaelic, has a rich and fascinating history dating back to the late Stone Age. Ranging from a Pictish circular stone broch, a 13th century cathedral and two ruined castles to disused 19th century lime workings, the island is full of historic sites. Following evictions and emigration during the 'clearances', the population of this fertile island has dropped from a high of 1,450 in 1845 to its present level of about 140, many of whom are still employed in the island's main industry of cattle and sheep raising.

Below *Until 1914, Port Ramsay, in the north of the island, was one of the five sites on Lismore where lime-burning was an important local industry. Today, many of the former crofters' and fishermen's cottages are let as holiday homes. Locally caught lobsters and farmed oysters are currently in high demand from top restaurants. Surprisingly, earth tremors are common on Lismore as the island straddles the Great Glen geological fault line which runs diagonally across Scotland.*

HISTORY

Evidence found on Lismore in the form of a stone axe head shows that the island has been occupied for at least 5,500 years. There are also many Bronze Age burial mounds and a well-preserved Iron Age circular stone broch at Tirfuir that was once occupied by Picts. Christianity came to Lismore in the 6th century with the arrival of the Irish missionary St Moluag, who established a small monastery on the island. However, the Vikings put an end to the monks' peaceful existence when they settled on Lismore in the 8th century.

Little more is known of Lismore until around the end of the 13th century, when the island was selected as the site for the cathedral of the Bishopric of Argyll, and it soon became an important religious centre for this part of Scotland. Part of the cathedral is now incorprated into the island's parish church.

Farming has always been, and still is, the most important industry on Lismore. For centuries the growing of cereal crops, along with sheep and cattle raising, were vital to the island's economy and, even today, cattle raised on Lismore are in high demand on the mainland. Lime-

burning was also an important industry on Lismore from the early 19th century until its demise in 1934.

Emigration to the New World started in the late 18th century, but by the time of the 'clearances' in the 1840s it had become a mass exodus. Many of the island's people, some forcibly evicted from their homes, sailed for a new life, not only to North America but also to Australia and New Zealand. From a peak of 1,450 in 1845, the population of the island continued to drop for many years until now, when it has stabilised at around 140. Today, although the raising of cattle and sheep are still the island's economical mainstay, tourism is also beginning to play an important role.

NATURAL HISTORY

It is not surprising that Lismore gets its name from the Gaelic *Lios Mor* or 'Great Garden'. Its fertile soil not only supports farming but also a large number of wild flowers, including primrose, yellow iris, meadowsweet and many species of rare orchid. In the skies can be seen buzzard, stonechat, hen harrier and skylark, while around the island's secluded and rocky coastline both seal and, with patience, the

occasional sea otter can be spotted going about their daily lives.

HOW TO GET THERE

By sea Caledonian MacBrayne operate a vehicle and passenger ferry between Oban and Achnacroish on Lismore. For more details contact their reservations office (tel. 08705 650000) or visit their website: www.calmac.co.uk

A passenger ferry also operates between Port Appin on the mainland and the northerly point of Lismore. For more details contact Argyll & Bute Council (tel. 01631 562125).

ORDNANCE SURVEY MAPS

Landranger 1:50,000 series No. 49

TOURIST INFORMATION

Nearest office: Oban Tourist Information Centre, Argyll Square, Oban, Argyll (tel. 01631 563122).
Website: www.oban.org.uk

WHERE TO STAY

There are several bed and breakfast and self-catering establishments on Lismore. For more details contact Oban Tourist Information Office (see above) or visit the island's website: www.isleoflismore.com

ISLAND WALKS

With very little traffic, the single-track roads of Lismore are a pleasant way to see most of the island. In the south, a rough

Above *Sailean, midway down the west coast of Lismore, was an important centre for the lime-burning industry until the local quarry closed in 1934. Today an eerie silence, occasionally punctuated by blasting from the Glensanda quarry on the mainland, has settled on the quarry, old lime kilns, disused harbour and the ghostly skeleton of the last of the 'smacks' that once shipped the lime.*

track continues where the road ends and leads to the most southerly point at Rubha Fiart. On the west coast, tracks lead to the disused lime workings and quay at Sailean and also to the ruins of Castle Coeffin.

SLATE ISLANDS: SEIL AND EASDALE

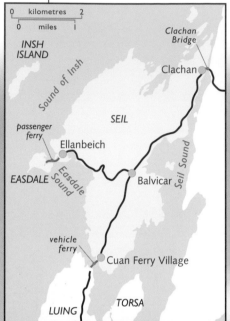

Although physically connected to the mainland since 1792, the island of Seil still clings to its island status. During the 18th and 19th centuries, slate quarries on both Seil and little Easdale Island exported millions of roofing tiles to the far-flung corners of the British Empire. Today the villages are silent, the quarries are flooded and the former quarry workers' houses are now holiday homes for the rich.

HISTORY

Christianity arrived on Seil during the 6th century after the Irish missionary, St Brendan, visited the island. The ruins of a chapel dating from this period can be seen today in Kilbrandon churchyard. Following Viking rule, Seil and Easdale Island became part of the kingdom of the Lord of the Isles until finally passing into the ownership of the Breadalbane Campbells in the 17th century. Because of their enormous deposits of slate rock Seil, with the islands of Belnahua, Luing and Easdale, became known as the Slate Islands and as early as the 16th century, slate from quarries on Easdale Island was being used to roof buildings and provide gravestones. By the early 19th century, millions of slate roofing tiles from quarries on Easdale and at Ellenbeich

Above and left *The only road access to Seil is via the Clachan Bridge which crosses the narrow Clachan Sound. Also known as the 'Bridge over the Atlantic', its 18th century stonework is covered with the pink flowers of the fairy foxglove during the summer. Once across the bridge, many visitors to Seil stop for refreshments at the Tigh-an-Truish pub. The name of the pub (House of Trousers) comes from the period following the 1745 Jacobite Rebellion when the wearing of kilts was banned in Scotland. Islanders changed into trousers at the pub before crossing to the mainland.*

were being exported annually around the world from the busy harbour at Easdale – during the period of peak production in the mid-19th century, the population of quarry workers and their families on tiny Easdale Island had risen to 450. Quarrying ceased at Ellenbeich in 1881 after a tidal wave engulfed the workings, but continued on Easdale Island until 1911. The Breadalbane Estate was broken up in the 1930s and, since then, Easdale Island has had a succession of private owners. The majority of the former quarrymen's cottages on the island are now holiday homes. On Seil today, the shellfish industry is an important economic activity along with farming and seasonal tourism.

NATURAL HISTORY

The islands of Seil and Easdale are located on the eastern edge of the Firth of Lorn, which was designated as a European Marine Special Area of Conservation in 2005. In particular, this area has been designated for its tide-swept rock reef habitats and species. Otters are regularly seen around the coastline, while seals are regular visitors to the flooded slate quarries on Easdale Island. On the island, toads and voles seem to survive happily in the old slate-walled vegetable gardens that were made from Irish soil imported during the 19th century. Once a hive of industry, the old quarry workings now provide an ideal habitat for many species of coastal wild flower while, in the

Above *Separated from Easdale Island by a narrow channel, the village of Ellenbeich is home to the Slate Islands Trust Heritage Centre which is housed in a former slate quarryworker's cottage. Slate quarrying at Ellenbeich ended in 1881 after a tidal wave engulfed the quarry.*

Below *With the Isle of Mull in the distance, this view of Easdale Island from Seil clearly shows the impact made on this tiny island by centuries of slate quarrying. By the early 19th century over five million roofing slates were being exported annually from the quarries on Easdale Island. Following flooding and increased competition from abroad, commercial slate quarrying on Easdale ceased in 1911.*

northeast corner of Seil, the cracks and crevices of Clachan Bridge, built in 1792, are an ideal habitat for the rare fairy foxglove (*Erinus alpinus*) which covers the bridge with its pink flowers during the summer. Created in the 1930s, the sheltered walled gardens of An Cala on the outskirts of Ellenbeich are open to the public between April and October.

HOW TO GET THERE
Seil: By road Take the B844 at Kilninver (on the A816, eight miles south of Oban) and follow to Clachan Bridge and Isle of Seil.
Easdale: By sea Argyll & Bute Council operate a regular passenger ferry from Ellenabeich on Seil to Easdale Island. For more details contact their area office in Oban (tel. 01631 562125).

ORDNANCE SURVEY MAPS
Landranger 1:50,000 series No. 55

TOURIST INFORMATION
Nearest office: Oban Tourist Information Centre, Argyll Square, Oban, Argyll (tel. 01631 563122). Website: www.oban.org.uk

WHERE TO STAY
Accommodation on Seil ranges from a hotel and and an inn to bed and breakfast and self-catering establishments. For more details contact Oban Tourist Information Centre (see above) or visit the island's website: www.seil.oban.ws Alternatively, visit the Easdale Island website: www.easdale.org

ISLAND WALKS
Seil Tracks and footpaths in the north of Seil lead from near the Tigh-na-Truish pub to Loch Caithlim and the ruined castle at Ardfad. From Kilbrandon Parish Church south of Balvicar, a track leads to Seil Sound. In the southeast, a track leads from tiny Port Mor past Ballachuan Loch

Right *Once the centre of a thriving slate industry, Easdale Island now slumbers on in blissful silence with many of its former quarrymen's cottages let as holiday homes Community life on the island centres around the Puffer Bar and the recently-built Community Centre.*

and across the B8003 to the Iron Age hillfort south of Dunmor House.
Easdale There are no roads or, apart from the odd wheelbarrow, any traffic on this small island. A short walk around the coastline takes in the old slate quarries, while the view from the top of the island's hill is well worth the climb.

Below *Slate, slate everywhere on Easdale Island. To the southwest, the tiny island of Belnahua was also an important exporter of slate roofing tiles until the outbreak of World War I in 1914. Now all that remains are a ruined village and flooded quarries.*

SLATE ISLANDS: LUING

For four centuries a small part of the Marquis of Breadalbane's estate, Luing and its slate quarries once supplied millions of roofing tiles to far flung corners of the British Empire. Since World War II, the island has again become justifiably world-famous – not only for its pedigree Luing breed of beef cattle, but also for its lobsters which are bred on the island's west coast within the EU-funded Firth of Lorn Special Area of Conservation.

HISTORY

Artefacts found on Luing show that the island has been occupied since at least the Bronze Age. Two hillforts, one overlooking Ardinamar Bay in the northeast and a very well-preserved example overlooking Lochan Iliter in the east, probably date from the Iron Age. Luing, along with the neighbouring islands of Shuna, Torsa, Seil and Easdale, became part of the kingdom of the Lord of the Isles until finally passing into the ownership of the Breadalbane Campbells in the 17th century. In addition to owning enormous tracts of land in Scotland, by the 18th century the Marquis of Breadalbane also owned the large island of Nova Scotia in Canada.

Because of their enormous deposits of slate rock Luing, with the islands of Belnahua, Seil and Easedale, became known as the Slate Islands and, by the 19th century, slate from quarries at Cullipool and Toberonochy was being shipped out from the pier at Black Mill Bay to many parts of the British Empire. Slate quarrying finally ended at Cullipool in 1966. The Breadalbane Estate was broken up in the 1930s and sold off as two farms. Since 1945, these farms have been owned by the Cadzow Brothers to raise sheep, beef and pedigree cattle. The world-famous Luing breed of beef cattle was started by the Cadzow Brothers in 1947 and became officially recognized as a true breed by the government in 1965.

Today farming, as well as clam, lobster and prawn fishing, are the island's main economic activities.

NATURAL HISTORY

Luing is located on the eastern edge of the Firth of Lorn, which was designated as a European Marine Special Area of Conservation in 2005. In particular, this area has been designated for its tide-swept rock reef habitats and species. Otters are regularly seen around the island's coastline while, out to sea, porpoise, dolphin and seal are regular visitors. On land are red deer, feral goats. many hares and, sadly, many mink. Along with its resident birds of prey such as peregrine, buzzard and hen harrier, Luing is also visited by many migratory birds during spring and

Right *By the 19th century millions of slate roofing tiles were being exported from this quarry at Cullipool to North America and far-flung parts of the British Empire.*

autumn. Up to 100 different species of native and migratory birds can be seen here throughout the year.

How To Get There
By sea Argyll & Bute Council operate a vehicle and passenger ferry between Cuan at the south of the island of Seil and Luing. For more details contact their area office in Oban (tel. 01631 562125).

Ordnance Survey Maps
Landranger 1:50,000 series No. 55

Tourist Information
Nearest office: Oban Tourist Information Centre, Argyll Square, Oban, Argyll (tel. 01631 563122). Website: www.oban.org.uk

Where to Stay
There are several self-catering establishments on Luing and one on the neighbouring island of Torsa. For more details contact Oban Tourist Information Office (see above) or visit the island's website: www.isleofluing.co.uk

Island Walks
With little traffic and about 10 miles of road on Luing, walking is one of the best ways to see the island. From Kilchattan, there are tracks to the southern tip of the island, to Black Mill Bay and to Lochan Iliter and Leacamor hillfort. At the north of Luing, footpaths and tracks connect Ardinamar with Cullipool (famed for its sunsets) and the abandoned village of Port Mary. Cycle hire is also available from Sunnybrae Caravan Park.

Above *The ruined chapel and cemetery at Kilchattan date back to the late 16th century. Although the chapel appears to have fallen out of use around 1735, the cemetery contains many interesting gravestones including that of Alexander Campbell, who died in 1787. The leader of an extremely intolerant and zealous religious sect called the Covenanters of Lorne, Campbell had expelled all of the other members by the time of his death. He even carved his own gravestone and dug his own grave prior to dying!*

Below *Overlooking the privately-owned island of Scarba, the disused pier at Black Mill Bay was once busy with the export of Luing roofing slate and steamer traffic to Glasgow. Where there was once a thriving village with its own shop and harbourmaster, today there is just this derelict office and the rusting remains of the pier.*

COLONSAY

Colonsay, or St Columba's Isle, is one of the most isolated communities in Great Britain. With its Bronze Age standing stones, Iron Age hillforts, early Christian chapels and links to the foundation of the Lord of the Isles, this Inner Hebridean island is rich in archæological and historic remains. Today, following several centuries of a declining population Colonsay, with its natural beauty, peace and tranquility, depends heavily for its economic survival on tourism.

Below *Exposed at low tide and fringed by a carpet of sea pink, The Strand is mile-wide stretch of sand that links Colonsay with its southern, smaller neighbour of Oronsay. Although used by vehicles (including the local post van), cyclists and pedestrians to reach Oronsay, care must always be taken to avoid being cut-off by the incoming tide.*

HISTORY

Archæological remains found on the island show that Colonsay has been occupied by humans for over 8,000 years. Standing stones found at several locations on the island such as at Machrins, Garvard and Lower Kilchattan date from the later Bronze Age, while the hillforts at Dun Meadhonach, Dun Eibhinn and Dun Cholla date from the Iron Age period, when the Dal Riada tribe from Ireland settled on the island. Following visits by the early Christian missionaries, including St Columba, Colonsay soon became an important Viking base. In a major battle off the coast of Colonsay in 1156 Somerled, the king of Kintyre, defeated a fleet led by his brother-in-law Goraidh mac Amhlaibh, becoming the first ruler, or Lord of the Isles, of the west coast and islands of Scotland. During medieval times, Colonsay was ruled by the chiefs of the McPhee clan from their stronghold in the former Iron Age fort at Dun Eibhinn, northwest of Scalasaig.

During the 'clearances' of the 18th and 19th centuries, many of the inhabitants of Colonsay were forcibly evicted from the island and sailed to find a new life in North America. Since 1841, the population of the island has fallen from nearly 1,000 to today's total of just over 100! Now, with visitors attracted by Colonsay's beautiful and tranquil landscape, tourism has replaced farming and fishing as the island's main economy.

NATURAL HISTORY

With its equable climate and wide range of habitats, ranging from rugged cliffs, sandy beaches and rock pools to moorland, woodland and lochs, Colonsay is a naturalists' paradise. Nearly 200 species of birds, including seabirds such as kittiwake, razorbill and guillemot as well as the golden eagle, corncrake, chough and overwintering geese have been recorded in recent years. Together with 500 recorded species of local flora, Colonsay has become a mecca for both birdwatchers and botanists alike. The gardens and

ORONSAY

Connected to the south of Colonsay at low-tide, the island of Oronsay is now farmed by the Royal Society for the Protection of Birds. A haven for the rare corncrake and chough, this peaceful island is a favourite haunt for birdwatchers. Oronsay is also rich in archæological and historical remains, including two shell mounds dating from the Middle Stone Age and an Iron Age hillfort on Dun Domhnuill. St Columba and his right-hand man, St Oran, are thought to have landed here on their way to Iona in the 6th century and monastic remains dating from that period have been found in the later Augustinian priory, of which substantial ruins remain today. Visitors to the site of the priory, located close to Oronsay Farm, can also see the medieval stone Oronsay Cross and many beautifully carved tombstones.

woodlands of Colonsay House, planted mainly in the 1930s, are famed for their extensive collection of rhododendron and exotic sub-tropical trees and shrubs from the southern hemisphere. Both the private gardens and woodlands of the house are open to the public.

How to Get There
By sea Caledonian MacBrayne operate a vehicle and passenger ferry between Oban and Colonsay. On Wednesdays during the summer, an additional ferry operates between Islay, Colonsay and Kennacraig. For more details contact their reservations office (tel. 08705 650000) or visit their website: www.calmac.co.uk
By air There are plans to introduce flights between Oban Airport and Colonsay, where a new runway and terminal building have been completed.

For further details contact Oban Airport (tel. 01631 710384).

Ordnance Survey Maps
Landranger 1:50,000 series No. 61

Tourist Information
Nearest office: Oban Tourist Information Centre, Argyll Square, Oban, Argyll (tel. 01631 563122).
Website: www.oban.org.uk
Alternatively, visit the island's website: www.colonsay.org.uk

Where to Stay
There is a wide range of accommodation available on Colonsay, ranging from the Colonsay Hotel to self-catering, bed and breakfast and a self-catering backpackers' hostel. For further details contact Oban Tourist Information Centre or visit the island's website (see above).

Above *In the northwest, the sweeping sandy crescent of Kiloran Bay is overlooked by Colonsay's highest peak, Carnan Eoin. Flanked by rocky outcrops and backed by sand dunes, this beautiful beach also receives the full force of Atlantic rollers.*

Island Walks
With its tranquility and natural beauty, Colonsay offers the walker a wide variety of natural attractions, ranging from craggy northern peaks, the sweeping sandy crescent of Kiloran Bay in the north west and the dramatic cliffs along the west coast, to inland lochs and the vast stretch of The Strand to the south, which links Colonsay to the island of Oronsay at low tide.

Bicycle hire is also available and, with only minimal traffic, Colonsay's 14 miles of single track roads are ideal for exploring the island by bike.

JURA

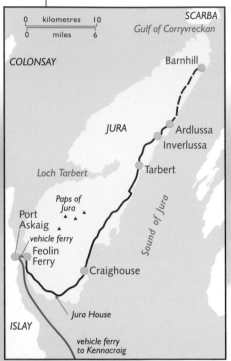

One of the last island wildernesses in Britain, Jura has been occupied by humans for at least 8,000 years. However, after centuries of Viking incursions and clan warfare, followed by the 'clearances' and a potato famine, most of the impoverished population of the island had emigrated to North America or Australia by the mid-19th century. Today, with a population of red deer that outnumbers humans by 30:1, Jura, with no direct link to the mainland, relies heavily for its economic survival on deer stalking, sport fishing, whisky production and tourism.

HISTORY

Archæological evidence, in the form of flint tools found near Inverlussa in the north of Jura, show that the island has been occupied by man for at least 8,000 years. These hunter-gatherers were then followed by Stone Age farmers, who lived on the island around 5,000 years ago. From the Bronze Age, there are numerous standing stones, cairns and hut circles to be found on Jura, while occupation during the Iron Age is represented by several forts such as An Dunan, which is set on a rocky outcrop overlooking the Sound of Jura near Lowlandman's Bay, four miles north of Craighouse.

There is little information on the history of Jura during this period or for the next thousand or so years, but it is known that there was a major battle on the island in the 8th century AD between the Britons and the Dalriada. From the 9th century Jura, claimed by both the King of Man and the King of Norway, was under constant attack from the Vikings. By the 12th century, however, the Vikings had been thrown out by Somerled who, by 1158, had become the military and political leader of the region. It is from Somerled's grandson, Donald, who later ruled Jura along with Islay during the early 13th century, that the clan Donald came about. They and their descendants, the MacDonalds, ruled the southern part of Jura until the 17th century, when they themselves were driven out by the Campbells. Meanwhile, the northern part of Jura was still in the hands of the MacLeans – and there was no love lost between them and the Campbells. There were several skirmishes between the two clans until the Campbells gained the upperhand; by the early 18th century, peace had settled on the island.

Left *Jura's population of pure-bred red deer outnumber the human population of around 180 by 30:1. Deer numbers are controlled during the stalking season which begins in July of each year.*

Apart from the laird and his cronies, life for the inhabitants of Jura was hard, with most of them scratching a living from the land or sea. Many emigrated to North America in the 18th century and even more left following the 'clearances' and potato famine of the 1840s. Peaking at 950 in 1851, Jura's population has continued to drop until today it numbers only around 180.

Over the years, the Campbells sold off their lands on Jura until the last member of the clan left just prior to World War II. Today, Jura is divided into eight separate estates and these, through their deer stalking and fishing activities, are the biggest employers on the island. Apart from this, the local economy is dependant on tourism, one whisky distillery and cattle and sheep raising.

Above *Cows wander back along the beach near Feolin Ferry after being milked. The Caol Ila distillery, just north of Port Askaig on Islay can be seen across the narrow sound that separates it from Jura.*

Below *Jura's only link with the outside world is this ferry, which is seen soon after leaving Feolin for the short crossing across the Sound of Islay to Port Askaig on Islay. Dominating the scene is the mountainous interior of this wild island - three of these peaks, known as the Paps of Jura, are over 2,500ft high. Originally shaped during the Ice Age, these quartzite peaks are unusually resistant to natural erosion.*

JURA HOUSE AND GARDENS

Located 3½ miles southwest of Craighouse, Jura House was built in the early 19th century by the Campbells and is the main house on the Ardfin Estate, which covers much of southern Jura. Its sheltered walled garden, warmed by the Gulf Stream, contains a large collection of plants from New Zealand and Australia that were planted at the beginning of this century. The gardens are open to the public all year and self-catering accommodation is also available in the house. For more details visit their website: www.jurahouseandgardens.co.uk

Above *Craighouse is the only large settlement on Jura and is home to the island's only distillery, hotel, shop, church and petrol pump. The one single-track road on Jura, built by Telford in 1812, leads from Feolin Ferry through Craighouse and up the east coast until it peters out two miles north of Ardlussa. From here it is a strenuous hike to the far north of island and the world-famous whirlpool of Corryveckan.*

NATURAL HISTORY

With its mountainous and treeless interior, its mainly inaccessible west coast and spectacular raised beaches, along with virtually no agricultural development and a tiny population, Jura is one of the last wildernesses in Britain. It not only supports a population of around 5,500 red deer but is also, due to its isolation, home to a large number of sea otter that live along its coastline. Both common and Atlantic grey seal can be seen basking in the bays and inlets, while out to sea are dolphin, porpoise and minke whale. Jura is also rich in birdlife including golden eagle, sea eagle, common and Arctic tern, peregrine falcon, chough and barn and tawny owl.

HOW TO GET THERE

By sea Argyll & Bute Council operate a vehicle and passenger ferry between Port Askaig on Islay to Feolin on Jura. For more details contact their agent's office in Port Askaig (tel. 01496 840681).

Caledonian MacBrayne operate a vehicle and passenger ferry between Kennacraig and Port Askaig/Port Ellen on Islay. For more details contact their reservations office (tel. 08705 650000) or visit their website: www.calmac.co.uk

By air British Airways currently operate two daily flights on weekdays and one flight on Saturdays and Sundays between Glasgow and Islay. For further details visit their website www.ba.com

Passengers wishing to travel to Jura must make separate arrangements for travel by bus or taxi between Islay Airport and Port Askaig.

ORDNANCE SURVEY MAPS

Landranger 1:50,000 series No. 61

GEORGE ORWELL

Eric Arthur Blair (1903–1950), better known by his pseudonym George Orwell, was born in India and educated at Eton. In 1922 he joined the Indian Police and served in Burma, but resigned after five years and became a writer. Following the publication of several books including *Keep the Aspidistra Flying* and *The Road to Wigan Pier*, he went to fight for the Republican side in the Spanish Civil War, in which he was seriously wounded. During World War II, Orwell worked for the BBC and was engaged in working on propaganda broadcasts to the Far East. After his critically acclaimed book *Animal Farm* was published in 1945, Orwell went to work as a journalist for *The Observer*, *Tribune* and *Manchester Evening News*. In 1946, wishing to find peace and solitude to write his next book and to escape life in London, Orwell moved to live on Jura. Over the next two years, he rented a remote farmhouse, Barnhill, in the north of the island, where he wrote his world-famous book *1984*. Already suffering from tubercolosis, Orwell only lived for a few months after its publication in 1949.

TOURIST INFORMATION

Nearest office: Bowmore Tourist
Information Centre, The Square, Bowmore,
Isle of Islay PA43 7JP (tel. 08707 200617)
or visit the VisitScotland website:
www.visitscotland.com

WHERE TO STAY

In addition to the 18-bedroom Jura
Hotel, there is also a good selection of
self-catering establishments and one
campsite on Jura. For more details
contact Bowmore Tourist Information
Centre (see above) or visit the island's
website: www.juradevelopment.co.uk

ISLAND WALKS

With only one road, a mountainous
interior and measuring nearly 30 miles
long and 8 miles wide, the wild but
beautiful island of Jura is a walkers'
paradise. Walkers are advised to be fully
prepared with suitable clothing and
footwear, OS map, compass and
provisions. On Jura, which is divided into
eight private estates, the deer stalking
period runs from 1 July to mid-February.
During this period, walkers should
contact the relevant estate keeper before
embarking on their journey.

Above *This eight-wheeled go-anywhere vehicle is the* de rigeur *form of transport in the wilderness and bogs of northern Jura.*

Below *High up on Jura's east coast is Lussa Bay, the mouth of the River Lussa and the tiny settlement of Inverlussa. It was close to here that archæologists once found thousands of flint tools that proved beyond doubt that Stone Age hunter-gatherer man once lived on the island over 8,000 years ago.*

ISLAY

For centuries the home of the mighty Clan MacDonald and power base of the Lords of the Isles, by the early 19th century Islay, the fifth largest island in Scotland, had fallen on hard times. Bankruptcy by its owners, famine and overpopulation all contributed to a mass exodus of islanders to pastures new in North America and the Antipodes. Today, the combination of improved transport links with the mainland, abundant wildlife, a rich historical heritage and, last but not least, nine malt whisky distilleries, all contribute to Islay's growing importance as a popular tourist destination.

Map

vehicle ferry to Colonsay (summer only)

Atlantic Ocean

Ardnave Point

Bunnahabbain

Sound of Islay

RSPB Centre

Loch Gruinart

Port Askaig

JURA

Finlaggan

Feolin Ferry

Loch Gorm

Machir Bay

Bruichladdich

Bridgend

vehicle ferry to Kennacraig

Port Charlotte

Loch Indaal

Bowmore

ISLAY

Mc Arthur's Head

Portnahaven

airport

ORSAY   Port Wemyss

Laggan Bay

Port Ellen

Ardbeg

Laphroaig

TEXA

The Oa

| 0 | kilometres | 10 |
| 0 | miles | 6 |

vehicle ferry to Kennacraig

HISTORY

Artefacts in the form of flint tools found on Islay show that the island has been inhabited for at least 10,000 years. From Bronze Age hut circles, field systems, burial cairns and standing stones to Iron Age hillforts, duns and crannogs, Islay is particularly rich in archæological remains. By about the 4th century AD, Islay had become part of the Gaelic kingdom of Dal Riada. The early Christian missionaries, who arrived from Ireland during the 6th century, left their mark in the form of several chapels and beautifully carved Celtic crosses, such as the fine examples at Kildalton and Kilnave, which can be seen on the island today. By the middle of the 9th century Islay, along with the rest of the Western Isles, had fallen into the hands of the

Left *Port Ellen is one of the two destinations on Islay served by the Caledonian MacBrayne ferry from Kennacraig. Sadly, the Port Ellen whisky distillery, which opened in 1825, ceased production in 1983. However, stocks of this sought-after single malt are still available.*

Below *The broad sandy sweep of Machir Bay on Islay's west coast is the final destination for these Atlantic rollers. Due west from here the next landfall is 2,500 miles away in Labrador, Canada.*

ISLAY MALT WHISKY

Characterised by its strong peaty flavour and dry finish, Islay malt whisky is famed the world over. Between them, Islay's nine distilleries probably produce more revenue for HM Government Exchequer each year than the GNP of a small third world country! Apart from the recently opened Kilchoman distillery which began production in 2005, all of the other distilleries are located on Islay's coastline: Bunnahabhain and Caol Ila in the northeast; Ardbeg, Laguvalin and Laphroaig in the southeast; and Bowmore, Bruichladdich and Port Charlotte around Loch Indaal in the west. Between April and October visitors to Islay can take part in guided tours of any of these distilleries. For more details visit the Islay Whisky Society's website: www.islaywhiskysociety.com

Vikings. Their hold on the region only ended when they were defeated by forces led by Somerled in a major sea battle off the coast of Islay in 1156. Known as the King of the Hebrides, Somerled ruled the region from Dunyvaig Castle in the south of the island. Its site, on a small headland overlooking Lagavulin Bay, can be seen today close to the Lagavulin malt whisky distillery.

Somerled died in battle at Renfrew in 1164 and his place as leader, or Lord of the Isles, was taken by his son Reginald. The latter died in 1207, to be succeeded by his son, Donald – founder of the immensely powerful Clan MacDonald. For the next 300 years, the Lords of the Isles ruled their kingdom from fortified islands set in Loch Finlaggan, a few miles inland from modern-day Port Askaig. One of the most important historical sites in Scotland, Finlaggan and its remains are now looked after by the Finlaggan Trust, who operate an interpretaive centre near the site.

Ownership of Islay passed to the Campbells of Cawdor in the early 17th century but, after years of clan feuding, peace only finally settled on the island when it was bought by the unrelated

Above *Built in 1815, the Ardbeg distillery is located on the south coast of Islay three miles east of Port Ellen. It was mothballed in 1981, but production re-started in 1997 with new maltings in Port Ellen. The original buildings at Ardbeg are still in use and house a café and shop.*

Daniell Campbell in 1726. Owner of the Shawfield estate near Glasgow, Campbell and his successors introduced many changes and breathed new life into the island. Under the Campbells, a brand new village with an unusual round church was built at Bowmore and the

influx of newcomers, attracted by the apparent success of the island, became so great that the population had risen to over 8,000 by 1801. By then farming, along with fishing, whisky and kelp production, were the economic mainstays of the islanders. By 1831, the population had increased to a staggering 15,000 but hard times were ahead for the stoic

Below *Probably occupied since at least the Iron Age, the crannog (or fortified island) of Eilean Mor at the northern end of Loch Finlaggan is one of the most historic sites in Scotland. It was from here that the Lords of the Isles ruled their kingdom between the 12th and 16th centuries.*

islanders. By this date emigration, which had started as a trickle in the late 18th century, suddenly turned into a flood. Bankruptcy by the Campbells, indebted tenant farmers, food shortages – partly caused by the failure of the potato harvests – and the later 'clearances', all led to a mass exodus for a new life in Canada, South Africa, Australia and New Zealand. By 1891, the population had halved to just over 7,000 and continued to decline into the 20th century, reaching rock bottom in 1991 with only 3,500.

In 1853, Islay was sold to James Morrison, one of the wealthiest men in Britain with estates as far away as North America. As a giant farm and sporting

estate, the island stayed more or less intact in the ownership of the Morrison family until the 20th century. Today, the traditional industries of farming and malt whisky production sit comfortably alongside forestry and tourism as the main economic activities on the island. In recent years, Islay's rich natural and historical heritage, together with improvements in transport links to the mainland, have helped to boost visitor numbers to this magnificent island.

NATURAL HISTORY

With over 270 species seen on the island, including over 100 that regularly breed here, Islay is a birdwatcher's paradise. Its

importance internationally has been recognised by the Royal Society for the Protection of Birds, who manage two large reserves on the island. In the north, the RSPB reserve at Loch Gruinart is one of the most important sites for overwintering barnacle and Greenland white-fronted geese and, during the spring, for wading birds such as snipe and lapwing. In the southwest, the rugged cliffs of The Oa, also an RSPB reserve, are home to the rare chough as well as guillemot, razorbill and raven. At both sites, birds of prey such as golden eagle, pergrine and hen harrier are regular visitors to the skies. For more details about RSPB activities on Islay, contact

their nature reserve office at Gruinart (tel. 01496 850505) or visit the RSPB website: www.rspb.org.uk/scotland

In addition to the two reserves, there are many other locations around the island which will delight the birdwatcher, including Loch Indaal, Bridgend Woods, Lochs Gorm and Skerrols, Machir Bay and The Rhinns.

Islay's rugged coastline and many sheltered bays are regularly visited by seals. Common seal can be seen on the southeast coast and in Loch Indaal, while the grey seal, which breed on little Nave Island in the north, can often be seen basking at low tide in Loch Gruinart and are regular visitors to the little harbour at

Above *For hundreds of years, the long and tortuous coastline of Islay has been a graveyard for ships driven ashore in Atlantic storms. The greatest tragedy of all was the sinking of the armed merchant cruiser HMS Otranto on 16 October 1918 - less than a month before the end of World War I - after a collision with HMS Kashmir in Machir Bay (seen above). 351 American soldiers and 80 British crew died.*

Below *The ancient chapel and 8th century Celtic cross at Kilnave stand on a hillside overlooking Loch Gruinart in the north of the island. Now a nature reserve managed by the RSPB, the head of the loch is also the site of a famous battle between the feuding MacLeans and MacDonalds in 1598.*

Portnahaven. Out to sea, there have been sightings in recent years of sperm, pilot, killer and minke whale in addition to dolphin and porpoise. The shy otter is fairly common on Islay and is found around the coast and inland lochs.

Several thousand red deer also inhabit Islay, mainly confined to the more inaccessible hills of the eastern half of the island. Other mammals likely to be seen on Islay are fallow and roe deer, feral goat – especially on The Oa – and the island's unique species of common shrew, *Sorex araneus granti* and field vole, *Microtus agrestis fiona*.

In recent years, over 20 species of butterfly and 12 species of dragonfly have been recorded on Islay. The island is also a botanist's paradise with over 900 plant species being recorded.

Below *On Islay's east coast, the harbour at tiny Port Askaig is usually a busy place with the comings and goings of the ferries to Feolin on Jura, Kennacraig on the mainland and the island of Colonsay (Summer Wednesdays only). The gardens of the nearby 400-year-old Port Askaig Hotel are a perfect spot to relax and enjoy views across the Sound of Islay to Jura.*

Founded in 1984, the Islay Natural History Trust runs an information centre with displays, library, lecture room, laboratory and sea aquarium at Port Charlotte (tel. 01496 850288). For full details visit the Trust's website: www.islaywildlife.freeserve.co.uk

HOW TO GET THERE

By sea Caledonian MacBrayne operate a vehicle and passenger ferry between Kennacraig on the Mull of Kintyre and Port Askaig/Port Ellen on Islay. For more details contact their reservations office (tel. 08705 650000) or visit their website: www.calmac.co.uk

Argyll & Bute Council operate a vehicle and passenger ferry between Port Feolin on Jura and Port Askaig on Islay. For more details contact their agent's office in Port Askaig (tel. 01496 840681).
By air British Airways currently operate two daily flights on weekdays and one flight on Saturdays and Sundays between Glasgow and Islay. For further details visit their website www.ba.com

ORDNANCE SURVEY MAPS
Landranger 1:50,000 series No. 60

TOURIST INFORMATION
Nearest office: Bowmore Tourist Information Centre, The Square, Bowmore, Isle of Islay PA43 7JP (tel. 08707 200617) or visit the VisitScotland website: www.visitscotland.com

WHERE TO STAY
There is a wide range of accommodation available on Islay. For more details contact Bowmore Tourist Information Centre (see above). Alternatively visit the island's three websites: www.islay.com, www.islay.co.uk or www.isle-of-islay.com

ISLAND WALKS
From the wild and impenetrable hillcountry and lochs of the southeast and the rocky shoreline of The Oa in the

Below *At the far western tip of the Rhinns of Islay is the fishing village of Portnahaven and its sheltered harbour. With fishing now in decline, many of the former fishermen's cottages are holiday homes. The world's first commercial wave-powered electricity generator started operation at Portnahaven in 2000. Scientists are now working on a plan to turn Islay into the world's first hydrogen powered island.*

southwest to birdwatching opportunities around Loch Gruinart in the north, Islay offers many different terrains and possibilities for the walker.

Between April and September, the RSPB organise guided walks at their reserves at Loch Gruinart and The Oa. For more information contact the RSPB office at Gruinart (tel. 01496 850505).

'Walkislay' is a walking week (guaranteed by the organisers to be midge-free!) held on Islay in April of each year. A programme of organised walks graded from easy through to strenuous includes visits to the neighbouring islands of Colonsay and Jura. For further information visit their website: www.walkislay.co.uk

Above *Loch Gruinart, here seen at low tide, is not only famous for its oysters but also as an internationally important site for overwintering barnacle geese. The Royal Society for the Protection of Birds manage both the Loch Gruinart Reserve, where there is a hide, and the reserve on The Oa, in the far southwest of Islay, where the windswept cliffs provide shelter for the rare chough.*

GIGHA

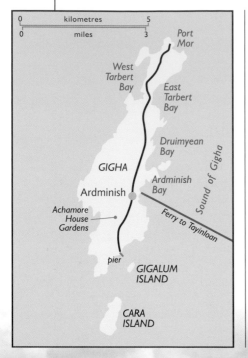

One of the most fertile islands in Scotland, little Gigha has seen its fair share of murderous clan feuding. Fought over for centuries by the MacNeills, MacLeans and MacDonalds and changing hands several times during the 19th and 20th centuries, the island's future was finally secured in 2002 when it was purchased by the islanders. Now run as a Trust, Gigha has successfully welcomed new businesses to help regenerate the island's community and economy.

HISTORY

Archæological evidence in the form of several standing stones, burial sites and a hillfort show that the fertile island of Gigha has been inhabited for at least 5,000 years. The early Christian missionaries from Ireland arrived in the 6th century, and remains of a chapel and an ancient cross from this period can be seen south of Tarbert in the north of the island. Following years of Viking influence - the name 'Gigha' comes from the Norse 'Gudey' or Good Isle – Christianity re-established itself in the 13th century, when a small church was built at Kilchattan in the south of the island. Now ruined, the church, which is dedicated to the 6th century Irish missionary St Catan, still contains its octagonal font and a large number of decorated graveslabs.

By the mid-15th century Gigha was owned by the MacNeills, but their rule was interrupted several times over the next 200 years following murderous attacks by both the MacLeans and MacDonalds. The MacNeills finally regained Gigha in 1631 and it remained in their hands until 1884, when it was sold to Captain William Scarlett. During the 20th century, the island changed hands several times until 2002, when it was purchased (with the help of grants) by the islanders, who now own it through the Isle of Gigha Heritage Trust.

NATURAL HISTORY

A wildlife haven, Gigha is renowned for both its flora and fauna. During early summer, the south of the island is ablaze

Below *With the Mull of Kintyre in the background, the CalMac vehicle and passenger ferry from Tayinloan approaches Ardminish on the Isle of Gigha. Now owned by the Isle of Gigha Heritage Trust, the island has recently seen an encouraging upsurge in its population, now standing at around 150, and growth of the local economy.*

ACHAMORE HOUSE & GARDENS

In the south of the island lie Achamore House and its world famous gardens. The house was built in 1884 by Captain William Scarlett, the 3rd Lord Abinger. To provide shelter, areas of woodland were also planted around the house at the same time. The estate was purchased in 1944 by Sir James Horlick, who was responsible for planting the 50 acres of gardens that can be seen today. The collection includes many rare sub-tropical trees and shrubs from South America, Asia, Australia and New Zealand and is particularly noted for its fine collection of rhododendron. The gardens are open to the public all year round. Bed and breakfast accommodation is available in Achamore House. For more details visit the house's website: www.achamorehouse.com

Right *The South Pier on Gigha is used for berthing purposes by local fishing boats for overnight mooring. It is also a good location for spotting seals. A small grass airstrip for private planes and microlights is located a short distance from the pier and close by are three wind turbines that provide electricity for the island and the National Grid.*

with a magnificent collection of wildflowers, including carpets of sea pink and bluebell. In addition to the collection of many rare sub-tropical trees and shrubs, the gardens at Achamore House (see below) also attract a number of bird species including tree creeper, great spotted woodpecker, long-tailed tit and yellowhammer.

The island's 30 miles of secluded coastline are a perfect place to see colonies of seal and the shy sea otter. To the east, the shallow Sound of Gigha provides winter shelter for large numbers of waterfowl, including the Slavonian grebe and long-tailed duck, and white-fronted and greylag geese. Diving gannet, dolphin and porpoise are also regular visitors to the seas around the island. The rocky islets offshore support colonies of tern that nest here during spring and summer.

HOW TO GET THERE
By sea Caledonian MacBrayne operate a frequent vehicle and passenger ferry between Tayinloan on the Mull of Kintyre and Ardminish on Gigha. For more details contact their reservations office (tel. 08705 650000) or visit their website: www.calmac.co.uk

ORDNANCE SURVEY MAPS
Landranger 1:50,000 series No. 62

TOURIST INFORMATION
Nearest office: Tarbert Tourist Information Centre, Harbour Street, Tarbert, Argyll & Bute PA29 6UD (tel. 08707 200624) or visit the Tarbert website: www.tarbert.info

Alternatively visit the island's website: www.gigha.org.uk

WHERE TO STAY
There is a surprisingly wide choice of accommodation on Gigha, ranging from the 13-bedroom Gigha Hotel to several bed and breakfast and self-catering establishment and a campsite. For more details visit the island website: www.gigha.org.uk

ISLAND WALKS
Only 7 miles long and 1 mile wide, this little island has only one road and negligible traffic. The Gigha Path Network Group have designated 12 off-road walks which are described in a locally available booklet. Cycle hire is also available from the island shop.

Below *At the northern end of Rothesay Bay on the Isle of Bute, the rotting remains of this old pier are a reminder of Port Bannatyne's important contribution to the war effort in World War II. Located at the entrance to Loch Striven, Port Bannatyne, with its floating dock, became the centre for marine salvage covering the whole of the Western Approaches.*

FIRTH OF CLYDE

Isle of Arran

Isle of Bute

Great Cumbrae

Ailsa Craig

ISLE OF ARRAN

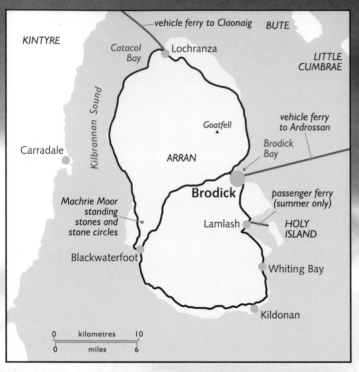

The largest island in the Firth of Clyde, the sixth largest in Scotland and the ninth largest around the coasts of Great Britain, the Isle of Arran is geologically located on the north-south divide of the Highland Boundary Fault. From its early history, when Arran was a sacred site for Bronze Age man, to Viking and Scots rule, Arran has for centuries been a microcosm of Scottish history. Its population, denuded by wholesale 'clearances' in the early 19th century and lack of job opportunities in the 20th century, is now on the upturn, thanks to improved transport links and the growth of the local tourist industry.

Right and below *Millions of years old, the granite peaks and glens of northern Arran have been shaped by ice, wind and rain. Both the surrounding glens and slopes of Goatfell, at 2,867ft the island's highest point, are now owned by the National Trust for Scotland. The lower forested slopes are managed by the Forestry Commission. Apart from boy racers on Brodick's seafront, the Isle of Arran is totally unspoilt and is a paradise for both walkers and nature lovers.*

HISTORY

The history of first inhabitants of the large and mountainous island of Arran is lost in the mists of time. However, they left behind substantial remains which can still be seen today on Machrie Moor in the southwest of the island. Here, one of the most important archæological sites in Scotland, the moorland landscape is littered with standing stones, hut circles and cairns, some of which date back to before the Bronze Age – over 4,000 years ago. Little is known about this early human habitation, but it is fairly certain that Machrie Moor was an important ritual site for an ancient culture that lasted for at least 2,000 years.

Not far from Machrie Moor, evidence of later occupation can be found in the Iron Age hill fort strategically located on the coast at Drumadoon. Later, a tribe of Celtic Britons, speaking a language akin to modern Welsh, may also have lived on Arran until they were replaced by the Dal Riada, a Gaelic-speaking people from northern Ireland, during the 6th century. It was during the 7th century that the first Christian visited Arran when St Mo Las arrived from Ireland. Rather than spreading the word, he spent years in solitary meditation in a cave on Holy Island in Lamlash Bay.

By the end of the 8th century Arran, along with the rest of the region, had became a target for Viking raids. Several artefacts dating from this period that have been found on the island also show that the invading Norsemen finally settled here. Their occupation of the island and the surrounding region only ended after their defeat at the hands of the Scottish king, Alexander III, at the Battle of Largs in 1263. Ownership of the Western Isles, including Arran, was finally settled at the Treaty of Perth in 1266 when they were sold by the Norwegian king, Magnus the Law-mender, to Scotland for 4,000 marks plus an annual annuity of 100 marks.

Now part of the kingdom of Scotland, Arran became the property of the Stewarts until it was given to the Hamiltons in the 16th century. During this period, Brodick Castle was attacked on many occasions, first by the English in 1351 and again in 1406, before being seized by the Duke of Argyll's forces who were supporters of Cromwell during the English Civil War.

Most of Arran was owned by the Dukes of Hamilton until the death of the 12th Duke in 1906. His only child, a daughter, subsequently became the Duchess of Montrose by marriage but, by her death in 1957, most of the lands had been either sold or, as in the case of Goat Fell, given to the National Trust for Scotland. The only exception was Brodick Castle, which also became the property of the National Trust in lieu of death duties.

After years of decline, Arran's population is now on the increase. Due to the 'clearances' of whole townships in the early 19th century to make way for more profitable sheep farming, thousands of islanders emigrated to Canada. From a peak of 6,600 in 1821, the population had fallen to 4,730 in 1881. Lack of job prospects brought further depletion in the

Left *Probably over 3,500 years old but with more recent graffiti, this slab of old red sandstone is one of many that litter the landscape of Machrie Moor in the southwest of Arran. Still covered by peat, the surrounding moorland has yet to give up many of its secrets from the Bronze Age.*

20th century and by 1971, Arran's population had dropped to an all-time low of 3,564. Fortunately, since then there has been a steady growth, no doubt helped by improved transport links and Arran's proximity to the main Scottish conurbations, and it currently stands at over 5,000. Today tourism, forestry and farming are the main economic activities on Arran.

NATURAL HISTORY

With a variety of landscapes, landforms and rocks, Arran is noted not only for its wildlife but also for its geology. As in mainland Scotland, the island is divided into two parts, highland and lowland, by the Highland Boundary Fault. In the late 18th century, the famous Scottish geolgist James Hutton (1726-1797) visited Arran to help prove his theory that granite was

Above *One of the most important archæological sites in Scotland, Machrie Moor is littered with hut circles, standing stones and cairns dating back to at least the Bronze Age.*

formed from the cooling of magma from deep in the Earth's crust. Through his findings on Arran and elsewhere in Scotland, Hutton became known as the father of modern geology.

Arran also has much to offer lovers of wildlife. The northern glens are not only home to around 2,000 red deer, but also provide shelter for well-camouflaged ptarmigan noted for their croaking call. In the skies, golden eagle, buzzard, peregrine, kestrel, sparrowhawk and hen harrier are regular sights.

In the south and west, Arran's mild climate supports many types of deciduous tree including two rare species of whitebeam unique to the island. The woods and forests around Brodick and in the south of Arran are also home to the shy red squirrel.

Also native to Arran are heron, shelduck, mallard, eider duck and merganser which are joined in the island's sheltered bays during winter by golden eye, teal and widgeon.

The waters around the Isle of Arran are regularly visited by diving gannet, basking shark, whale, dolphin and porpoise, while along its long coastline, seal and otter can also be seen. Holy Island (see below), reached by ferry from Lamalash, is now a wildlife haven protected by its environmentally friendly Buddhist owners.

HOW TO GET THERE

By sea Caledonian MacBrayne operate vehicle and passenger ferries between Ardrosssan and Brodick (all year) and between Claonaig on the Mull of

Right In the far north of Arran, the sheltered anchorage of Loch Ranza is a favourite haunt of yachtsmen. Now just a ruin, Lochranza Castle dates back to the 13th century when it was owned by the MacSweens. It later became a royal castle and was given by James II to Lord Montgomery. From Lochranza there is a useful vehicle ferry link to Claonaig at the head of the Mull of Kintyre. The village is also home to the island's only malt whisky distillery, which opened as recently as 1995.

Kintyre and Lochranza (seasonal). For more details contact their reservations office (tel. 08705 650000) or visit their website: www.calmac.co.uk

ORDNANCE SURVEY MAPS

Landranger 1:50,000 series No. 69

TOURIST INFORMATION

Nearest office: Brodick Tourist Information Centre, The Pier, Brodick, Isle of Arran, Strathclyde KA27 8AU (tel. 01770 302140/302401) or visit website: www.ayrshire-arran.com

WHERE TO STAY

There is a wide range of accommodation available on Arran. For more details contact Brodick Tourist Information Centre (see above) or visit the island website: www.visit-isle-of-arran.eu

ISLAND WALKS

Within reason, the whole of Arran is accessible on foot. Almost 60 miles in circumference and with only one major (coastal) road, the island offers many walking opportunities. The northern 'highland' half of the island is wild, mountainous and remote while the

HOLY ISLAND

Located in Lamlash Bay off the southeast coast of Arran, little Holy Island has been a place of religious pilgrimage for nearly 1,500 years. In the 7th century the early Christian monk St Mo Las arrived from Ireland and lived for years in a cave on the island in solitary meditation. A monastery was established here in the 13th century but had fallen out of use by the late 16th century. In 1991, the island was bought by a group of Scottish Buddhists who have worked to conserve the island's unique species of plants, protect the resident wild Eriskay ponies, Soay sheep and goats, and planted 35,000 native hardwood trees. Visitors wishing to travel to the island should first contact the Holy Island Ferry (01770 600998) or visit the island's website: www.holyisland.org

southern 'lowland' half is mainly forested, with much of the land owned by the Forestry Commission.

In addition to the numerous forest trails in the south, some of the best walking can be found among the wild northern glens and peaks. In particular, Arran's highest point, Goatfell, offers panoramic views on a clear day of the southwest of Scotland. In the southwest, the area north of Blackwaterfoot offers excellent walking opportunities to visit a whole range of historic sites, from standing stones and hut circles to the famous King's Cave where, legend has it, Robert the Bruce once lived.

Below *Overlooking Kilbranan Sound and just around the headland from Lochranza lies the broad sweep of Catacol Bay. In the tiny village of Catacol, a row of 12 white-painted terraced cottages, known as the 'Twelve Apostles', date back to the 1860s when they were built to house local people forced out of nearby Glen Catacol during the 'clearances'.*

ISLE OF BUTE

From Stone Age man and early Christian missionaries to marauding Vikings, battling Scots and the cruel English, the fertile and sheltered island of Bute has a long and fascinating history dating back at least 6,000 years. During the 19th century its capital, the Royal Burgh of Rothesay, became an important centre for the cotton-spinning and weaving industry before becoming a paradise for wealthy Glaswegian merchants and holidaymakers. Today, dairy and beef farming, together with forestry and tourism, are the island's main economic activities.

HISTORY

Flint tools found near Kilchattan Bay show that the fertile and sheltered island of Bute has been inhabited for at least 6,000 years. These early hunter-gathers were replaced about 5,000 years ago by the first farmers on the island who left behind several burial chambers, or cairns, which can be seen today along Bute's steep northwest coastline. Both the earlier flint tools and artefacts from a domestic site of the later period, unearthed in Rothesay, can be seen today in the town's museum. Also to be seen in the museum are artefacts unearthed from the many Bronze Age burial mounds, or cists, that are scatterered around the island – one of the best examples can be seen near Scolpsie Bay in the southwest – and the later Iron Age hillforts, of which the best example is at Port Dornach in the far south.

The early Christians from Ireland arrived on Bute in the 6th century and built a monastery on the site of what is now St Blane's church, in the south of the island. There are also the remains of two chapels from this period: St Ninian's in the west and St Maccaile's in the north. From the 8th century onwards, however, the peace and tranquility of Bute became increasingly disturbed by Viking incursions and it was not until 1266 that Bute finally became part of Scotland. In the intervening years, much blood was shed in battles between the

Right *The beached wreck of the fishing vessel* Co-Worker, *which dragged its anchor and ran aground in Ettrick Bay in 2000, is now one of Bute's more unusual attractions. In the misty distance is the island of Inchmarnock and, beyond, the Isle of Arran. From 1905 until its closure in 1936, an electric tramway linked Ettrick Bay to Rothesay.*

INCHMARNOCK

Two miles long and half a mile wide, the privately-owned island of Inchmarnock is located off the southwest coast of Bute. The island contains several sites of historic interest, including a Bronze Age stone coffin and the remains of a 7th century chapel established by the early Christian missonary, St Marnoc. Once divided into three farms, Inchmarnock was used as a base for commando training during World War II. Following the end of the war, the island remained uninhabited for over 30 years until it was purchased by its present owner, Sir Robert Smith, in 1999. It is now run as an organic farm with a herd of Highland cattle.

Scots and the Norsemen: the Isle of Bute and its stronghold, Rothesay Castle, changed hands several times. In 1335, during the wars of independence from England, Bute again came under fierce attack, this time from the massed forces of Edward III of England.

In 1398, Robert III of Scotland created the title of the Duke of Rothesay, and from this date Rothesay Castle became their main residence. Robert's son David Stewart, heir to the Scottish throne, became the first duke. Since then, the heir to the Scottish (and later, after the union of both countries in 1603, the

heir to the English) throne has always been given this title.

During the English Civil War, the Sheriff of Bute, Sir James Stewart, sided with the Royalists. In response to this, in 1650, Parliamentarian troops seized Rothesay Castle and held it for nine years. They partially destroyed the castle when they left, and its destruction was finally completed 30 years later when its remains were consumed by fire, after the town was attacked by the Earl of Argyll during his rebellion against James VII.

The industrial revolution arrived on Bute in 1779 when a cotton mill was

Above *Located close to the northern entrance of the Mount Stuart Estate, the model village at Kerrycroy was built in 1803 by the 1st Marquess of Bute. The two mock-Tudor houses were added in the 1890s. Until the pier at Rothesay was built in 1822, the little harbour at Kerrycroy was the point of departure for a ferry service from to the Ayrshire coast.*

Above *In a prominent position overlooking Rothesay Bay, the Glenburn Hotel was originally built as the Glenburn Hydropathic in 1843. Its cold water baths were seen as a health cure by wealthy Glaswegians. The hotel with its terraced gardens was rebuilt in its present form at the end of the 19th century*

established in Rothesay. The cotton industry expanded in the town until the mid-19th century, when there were three spinning and weaving mills employing over 1,000 people. However, faced with increased competition from mainland mills, the industry had completely disappeared from the island before the end of the century. By the mid-19th century, the opening of railways in Scotland and better links with the mainland quickly made Bute an attractive

and healthy location for wealthy Glaswegian businessmen and their families. Virtually all of the coastline, from Bannatyne through Rothesay and down to Ascog, became a building site for the opulent mansions and hotels that can still be seen today. However, the *pièce de resistance* from this period must surely be the flamboyant Gothic revival edifice built by the 3rd Marquess of Bute at Mount Stuart, just south of Ascog. The house, with its new Visitor Centre and world-famous gardens, is open to the public. For more details visit the house website: www.mountstuart.com

From the late 19th century Bute, and in particular Rothesay, became a popular destination for holidaymakers wishing to escape the 'dark satanic mills' of Glasgow. A thriving industry soon grew to cope

with this massive influx of summer visitors – by the late 1920s, around one million people a year were visiting Rothesay, where hotels, boarding houses, health spas, a bandstand, Winter Gardens and dance hall catered for their every need! The onset of World War II brought an end to this frivolity, and for the duration Rothesay, strategically located in the Firth of Clyde, became an important submarine base.

Until the end of the 19th century, cotton-spinning and weaving, fishing,

Below *Built of cast iron and glass in 1924, when Rothesay was still a popular holiday destination for Glaswegians, the domed Winter Gardens is now a Grade A listed building which currently houses a small cinema, restaurant and tourist information centre.*

Below *CalMac operate a reular vehicle and passenger ferry between Rothesay and Wemyss Bay where there is an excellent connection with train services to Glasgow.*

farming and boatbuilding were the mainstay of the island's economy. Today, the main economic activities of the 7,000 or so islanders are forestry, dairy, beef and sheep farming and tourism. The latter went into decline after World War II, but, although visitor numbers have never recovered to their halcyon days, the future is much brighter.

Natural History

Divided into two by the Highland Boundary faultline, the Isle of Bute is famed for its wide range of natural habitats and extraordinary diversity of wildlife. North of the faultline, the hillier northern part of the island, with its moorland and woodland, is particularly rich in birdlife and is also a favourite location for spotting red deer, hare and wild goat. Located on the faultline in the centre of the island, Loch Fad is famous for its trout fishing and is classified by Scottish Natural Heritage as a site of Special Scientific Interest for its scenic beauty and its huge diversity of bird and plant life. On the west coast, Scolpsie Bay, also directly on the faultline, is home to a colony of around 200 seal. By contrast, the more cultivated southern half of the island, with its rugged coastline, is taken over mainly by farming and forestry.

For the botanist, Bute has much to offer, including the famous Ascog Hall Fernery, south of Rothesay, which houses about 80 sub-tropical species of fern from the southern hemisphere. Also of interest nearby is the walled garden of Ardencraig House which is now owned by Argyll & Bute Council.

How to Get There

By sea Caledonian MacBrayne operate two vehicle and passenger ferry services to Bute. One runs from Wemyss Bay to Rothesay, the other from Colintraive on the Cowal peninsula to Rhubodach in the north of the island. For more details contact their reservations office (tel. 08705 650000) or visit their website: www.calmac.co.uk

Ordnance Survey Maps

Landranger 1:50,000 series No. 63

Tourist Information

Nearest office: Isle of Bute Discovery Centre, Winter Garden, Rothesay, Isle of Bute, Strathclyde PA20 0AT (tel. 08707 200619). Alternatively, visit the island's website: www.visitbute.com

Where to Stay

There is a wide choice of accommodation on Bute, ranging from luxury hotels and bed and breakfast establishments to self-catering properties and a campsite. For more details contact the Isle of Bute Discovery Centre (see above) or visit the island's website: www.visitbute.com

Island Walks

With a total mileage of 29 miles, the West Island Way runs the length of Bute. Starting in the south at Kilchattan Bay, the walk takes in St Blane's Church, Loch Fad and Glen More before ending close to the ferry terminal at Rhubodach.

St Blane's Church

Located on a remote hillside in the far south of the island of Bute, St Blane's church is also the site of a Celtic monastery that was established by St Catan in the 6th century. The church is dedicated to St Blane, Catan's nephew, who was born on Bute at that time. The original monastery was destroyed by the Vikings at the end of the 8th century and then abandoned. A rectangular building in the lower burial ground, remains of monks' cells and the enclosure wall all probably date from that early period. The present roofless building with its fine stonework (right) dates from the 12th century and contains many early gravestones. It is likely that the church fell out of use at the end of the 16th century.

GREAT CUMBRAE

O nce the base for an enormous Norwegian fleet before the inconclusive Battle of Largs in 1263, Great Cumbrae slumbered on until the 19th century when its main town, Millport, was developed into a popular seaside resort. Today the island, the most densely populated in Scotland, with its perfectly preserved Victorian town, tiny cathedral, university marine biological station and panoramic views over the Firth of Clyde, is a popular destination for daytrippers and cyclists.

Below *A popular holiday destination during the 19th century, Millport is still a perfectly preserved Victorian seaside town. Its cathedral, the smallest in Europe, was designed by William Butterfield in 1851.*

HISTORY

Great Cumbrae was probably used as a base by the Norwegians in their war against the Scots during the 13th century. Following a series of fruitless peace talks, a large Norwegian fleet under King Haakon IV sailed up the Firth of Clyde at the end of September 1263 and anchored off the island. However, several of his ships were driven ashore on the mainland during a storm, and Haakon landed a party of his men to protect them. The ensuing Battle of Largs was inconclusive, even though the Norwegians were vastly outnumbered by the Scots. Both parties withdrew and the Norwegians sailed back home. Haakon died in the Orkneys and ownership of the Western Isles, including Great Cumbrae, was only finally settled at the Treaty of Perth in 1266 when they were sold by Haakon's successor, Magnus the Law-mender, to Scotland for 4,000 marks plus an annual annuity of 100 marks.

For centuries afterwards, ownership of Great Cumbrae was split between the Earl of Glasgow, who owned the east, and the Marquess of Bute, who owned the west. The island was finally sold to its tenant farmers in 1999 by the last feudal owner, former racing driver Johnny Dumfries (the 7th Marquess of Bute).

Easily accessible from the mainland, the island's only town, Millport, with its sheltered sandy beaches, became a popular holiday destination for Glaswegians during the 19th century. Following scientific research by the Victorian naturalist David Robertson, a floating marine laboratory was established near Millport in 1897. The laboratory and aquarium, open to the public, are now run jointly by the Universities of Glasgow and London.

NATURAL HISTORY

For such a small island, Great Cumbrae is particularly rich in wildlife. Its shores teem with many bird species, while inland wild flowers, including nine species of orchid, and ferns abound. The Robertson Museum & Aquarium, based at Keppel Pier in the southeast, houses an important collection of marine life found in the waters around the island.

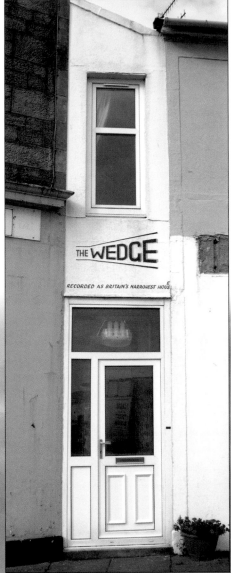

HOW TO GET THERE
By sea Caledonian MacBrayne operate a vehicle and passenger ferry between Largs and Cumbrae Slip. For more details contact their reservations office (tel. 08705 650000) or visit their website: www.calmac.co.uk

ORDNANCE SURVEY MAPS
Landranger 1:50,000 series No. 63

TOURIST INFORMATION
Nearest office: Largs Tourist Information Centre, Railway Station, Main Street, Largs, Strathclyde (tel. 01475 673765). Website: www.ayrshire-arran.com

WHERE TO STAY
There are several hotels, bed and breakfast and self-catering establishments

Above *Overlooking sheltered Millport Bay, the war memorial on Millport's seafront honours the island's fallen heroes from both World Wars.*

and a caravan site on Great Cumbrae. For more details contact Largs Tourist Information Centre (see above) or visit the island's website: www.millport.org.uk

ISLAND WALKS
Only 4 miles long and 2 miles wide, Great Cumbrae can be easily walked around in a day. With many daytrippers, however, the island's coastal road can get busy. The view of the Firth of Clyde from the island's highest point at the Glaid Stone (416ft) is well worth the climb. Cycling is a very popular option for getting round the island and there are several cycle hire shops in Millport.

Above *Millport not only boasts that it is home to Europe's smallest cathedral, but it also lays claim to the narrowest house in Britain. At only 47 inches wide there can hardly be any room to swing a cat!*

LITTLE CUMBRAE

To the south of Great Cumbrae is the privately owned island of Little Cumbrae. A religious cell was established here by St Veya in the 6th century and the small castle, which dates from the 13th century, was destroyed by Parliamentary forces during the English Civil War. Over the centuries, Little Cumbrae was populated by tenant farmers and owned at various times by the Hunters of Hunterston, the Montgomerys and the Earls of Eglington. On the west coast, the old lighthouse was built in 1793. On the east coast, the mansion house was built in 1913 with gardens planned by Gertrude Jekyll. In 2007 this 684–acre island, including a 12 bedroom mansion house, boathouse, jetty, two cottages, the 13th century keep and lighthouse complex with three houses was up for sale for £2.5 million. The island has a wide range of plant species together with grassland, bracken and heather. A recent survey by the Royal Society for the Protection of Birds listed 57 species of bird on the island.

AILSA CRAIG

Looking like a half-submerged prehistoric monster, the granite outcrop of Ailsa Craig is easily recognisable from the Ayrshire coast. Once a provider of food for the mainland and a shelter for fishermen, the island has for centuries been owned by the Earls of Cassilis. Now a bird sanctuary, Ailsa Craig is renowned not only for its unique granite used in the making of curling stones but also for its seabirds, including one of the largest gannet colonies in the world.

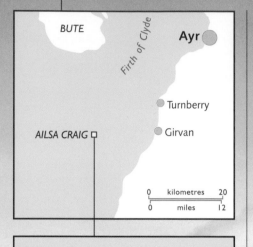

HISTORY

For centuries, Ailsa Craig not only provided shelter for local fishermen, but also provided an abundant amount of seabirds for mainland consumption. The island was first granted by Robert III to the Abbot of Crossraguel Monastery in 1404. It is likely that a small chapel was built on the west side of the island around this time, but its site was destroyed in the 1880s during construction of the new lighthouse.

By the mid-16th century, Ailsa Craig was owned by the Earls of Cassilis who lived at Turnberry Castle on the nearby mainland, and it is thought that the small Ailsa Castle was built around this time. However, following the Reformation and the threat of invasion by Spain, there were concerns that Ailsa Craig was about to be seized by Catholic forces and in 1597, to deter any landings, the Earl of Cassilis entrusted the safety of the island to Thomas Hamilton. The latter obviously took his job seriously, as his coat of arms can still be seen on the south wall of the castle today. Fortunately, Ailsa was not invaded and, for centuries, it continued its peaceful role as a provider of food – rabbits were also introduced to the island – and as a haven for fishermen.

Formed during a major volcanic eruption over 500 million years ago, Ailsa Craig is famed for its geology and its links with the sport of curling. Its unique granite, also known as Ailsite, was found to be the perfect material for producing highly polished curling stones during the 19th century. Soon, a thriving quarrying business became established on the island, with stones being exported around the world. Today the quarrymen's cottages are derelict, but granite from Ailsa is still used by curling stone manufactuers on the mainland.

Although clearly visible for miles on a clear day, Ailsa's location in the middle of a busy shipping lane contributed to several shipwrecks on the island during times of fog. In 1882, the Northern Lighouse Commissioners purchased land on the island to build a lighthouse and foghorns. Supervised by Thomas and David Stevenson, the lighthouse, associated buildings and two foghorns were completed in 1886. The foghorns, one at the south and one at the north of the island, were powered by gas blown along an underground pipe by compressed air. Together with the resident quarrymen, the four lighthouse keepers and their families had boosted the island's population to its peak of 29 by 1891. Until wireless communication was introduced in the 1930s, the only means that islanders had to communicate with the mainland was by carrier

pigeon or, in bad weather, by lighting fires as signals. By 1990, however, when the lighthouse became automated and with the quarrying operations long since shut down, the island's human population had dwindled to zero. Still owned by the Cassilis Estates, Ailsa Craig is now a bird sanctuary managed by the Royal Society for the Protection of Birds.

NATURAL HISTORY

With its steep-sided cliffs rising, in places, almost vertically from the sea, Ailsa Craig is first and foremost a breeding ground for seabirds. Once hunted intensively for their eggs, meat and feathers for consumption on the mainland, their much-reduced numbers are now protected by

Above *Located on a flat spit of land on the island's east coast, the Ailsa Craig lighouse was completed by Thomas and David Stevenson in 1886. Its oil-burning light was converted to electricity in 1911 and then to solar power in 2001. Now fully automated, the lighthouse is only visited by maintenance men flown in from the mainland by helicopter.*

the Royal Society for the Protection of Birds. Sadly, the puffin population, which once numbered several hundred thousand pairs, was decimated by brown rats which were accidently introduced in the late 19th century.

Although the rats have recently been eradicated, the return of the likeable puffin in any large quantities is still awaited. Ailsa Craig is also designated as a

Left *Even on a clear day, Ailsa Craig often has a tea-cosy of cloud on its 1,114ft-high summit. Now managed as a bird sanctuary by the RSPB, Ailsa Craig is world-famed not only for the curling stones made from its unique granite but also, during the breeding season, for its enormous colony of gannets. Apart from visitors, the island is now devoid of any human habitation.*

Below *In foggy conditions, the enormous bulk of Ailsa Craig proved to be the graveyard of many ships sailing up the Firth of Clyde. To warn shipping of the danger, a lighthouse and two foghorns were commissioned in 1886. Silent since 1966, this foghorn at the south end of the island was originally powered by gas blown by compressed air along an underground pipe. During foggy weather, the foghorn sounded alternating high and low notes every three minutes.*

Special Protection Area for Birds, and its steep-sided cliffs are home to one of the largest gannet colonies in the world. Unmolested by the brown rat, this colony now numbers around 70,000 pairs during the breeding season. Other seabirds that can be seen around the island are guillemot, razorbill, kittiwake, cormorant and herring and black-backed gull.

Introduced as a source of food in the 19th century, rabbits and Soay sheep still populate the island's slopes, but the badgers and racoons introduced by Lord Ailsa in the late 19th century are no more. Although virtually treeless, Ailsa's plant life is more abundant with fine displays of wild hyacinth, red campion and the rarer tree mallow in the summer.

HOW TO GET THERE

By sea Weather permitting, Mark McCrindle (tel. 01465 713219) is licensed to operate daily sailings from Girvan Harbour to Ailsa Craig all year round. For more details visit his website: www.ailsacraig.org.uk

ORDNANCE SURVEY MAPS
Landranger 1:50,000 series
No. 76

TOURIST INFORMATION

Nearest office: Girvan Visitor Centre, Bridge Street, Girvan, Strathclyde KA26 9HH (tel. 01465 715500).

WHERE TO STAY

There is no accommodation on Ailsa Craig. Accommodation is available on the mainland in Girvan and the surrounding area of south Ayrshire. For more details contact Girvan Visitor Centre or visit the website: www.visitscotland.com

ISLAND WALKS

Walking on Ailsa Craig is not recommended for the fainthearted. The island is also an RSPB reserve so walkers should keep to tracks or footpaths. A steep zig-zag path leads from just south of the lighthouse to the summit (1,114ft) passing the remains of the 16th century castle on the way. Depending on the tide and with care it is also possible to walk around the 2-mile long coastline of the island. Caution: beware of precipitous cliffs and occasional rock falls.

Above *Visitors from Girvan to Ailsa Craig are landed at this small jetty on the east coast. Once populated by quarrymen, lighthouse keepers and their families, the island has a certain* Marie Celeste *quality about it. From an overgrown narrow gauge railway and its wagons, to a winch engine, mountains of waste granite, rusting foghorns and ghostly, roofless and deserted buildings, their tangible remains are scattered all around the landscape. The eerie silence is only punctuated by the call of seabirds and the barking of seals.*

INDEX

AUTHOR'S ACKNOWLEDGEMENTS

I would like to thank the many people who have made the production of this book possible: to the publishers, Frances Lincoln, and in particular to John Nicholl; to the many seafarers who carried me safely to my destinations including the cheerful and helpful staff of Caledonian MacBrayne, Northlink Ferries, Orkney Ferries and Shetland Ferries; to Angus Campbell of Kilda Cruises and to Mark McCrindle of Girvan; to the staff and pilots of Directflight at Tingwall Airport on Shetland; to VisitScotland for their efficient and friendly service in arranging accommodation; to the cheerful and obliging natives of all of the Scottish islands; to the Scottish weather for being extremely accommodating; to all of my many friends and supporters (too numerous to mention) in Glastonbury, but in particular to John and Brigid of Beckets; to my wife Sarah for encouragement and friendship during many of my trips; and, last but not least, to Miranda Smith, Denise Stobie and Chas Stoddard, for their enormous contribution and unstinting encouragement during the latter stages of the production of this book.